W9-AVT-392

Time Journeys

Time Journeys

A Search for Cosmic Destiny and Meaning

Paul Halpern

Assistant Professor of Mathematics and Physics
Philadelphia College of Pharmacy and Science

McGraw-Hill, Inc.

New York   St. Louis   San Francisco   Auckland   Bogotá
Caracas   Hamburg   Lisbon   London   Madrid
Mexico   Milan   Montreal   New Delhi   Paris
San Juan   São Paulo   Singapore
Sydney   Tokyo   Toronto

Library of Congress Cataloging-in-Publication Data

Halpern, Paul.
    Time journeys: a search for cosmic destiny and meaning/Paul Halpern.
        p.   cm.
    Includes bibliographical references.
    ISBN 0-07-025704-3   —   ISBN 0-07-025706-X (pbk.)
    1. Time.   2. Physics—Philosophy.   3. Cosmology.   I. Title.
QB209.H25   1990
529—dc20   90-30141
                                                                   CIP

The author gratefully acknowledges permission to use passages from the following works.

Jorge Luis Borges: *Labyrinths.* Copyright © 1962, 1968 by New Directions Publishing Corporation. Reprinted by permission of New Directions Publishing Corporation.

"Lament" from *The Glass Bead Game* by Hermann Hesse. English translation © 1969 by Holt, Rinehart and Winston, Inc. Reprinted by permission of Henry Holt and Company, Inc.

    2 3 4 5 6 7 8 9 0   DOC/DOC   9 8 7 6 5 4 3 2 1 0

ISBN   0-07-025704-3
ISBN   0-07-025706-X {PBK.}

The sponsoring editor for this book was Jennifer Mitchell, the editing supervisor was Jim Halston, the designer was Naomi Auerbach, and the production supervisor was Suzanne W. Babeuf. It was set in Baskerville by McGraw Hill's Professional & Reference Division composition unit.

Printed and bound by R. R. Donnelley and Sons.

For more information about other McGraw-Hill materials, call 1-800-2-MCGRAW in the United States. In other countries, call your nearest McGraw-Hill office.

Dedicated to my grandmother, Esther Halpern, on the occasion of her eighty-fifth birthday.

Contents

Preface

Time affects us all, yet it is truly understood by none. Like the mythical Janus or the Hindu Lord Brahma, time presents us with many faces. It is both a creator and a destroyer; it brings us both hope and horror. Sometimes we reap its harvest; at other times we suffer its famine.

How can we describe time? How can we define it? Can we ever hope to catch and question this extremely powerful but ever-elusive demon? Or are we doomed to watch it slip through our hands, controlling us as it evades our grip? Our fate, in that case, would be to continue serving as prey to something that we could never hope to understand.

Can we think of time simply as a malevolent demon, bringing death and hardship, ruining all that it touches with its cold, unyielding grasp? Perhaps, on the other hand, it is more like a benevolent sprite, mysteriously providing us with hope and fortune. Or maybe, in fact, time is a blind god, striking down friends and foes alike as it sways its scythe to and fro without mercy or malice, rhyme or reason, as it staggers about.

Demons, sprites, gods—these are all "things"; they represent the tangible. Historically, humankind has sought to represent the unknown as something real and imaginable. Therefore, time has been thought of as "creator," "preserver," or "destroyer," and sometimes even as a combination of all three things as in the Hindu concept of Brahma. Yet our experience tells us that time is not a thing: it somehow exists outside the world of things.

The essence of time lies precisely in the fact that it transcends, yet defines, the realm of the mundane. Every worldly happening can only be seen as taking place within a temporal framework, yet, paradoxically, this framework exists only because events *do* happen. The world would not exist without time, and time would not exist without the world.

Thus, when we talk about different temporal outlooks, we really must refer to different ways of interpreting the history of the world or of describing the events of our lives.

Even the most ordinary occurrences, things that we take for granted, reveal the complexity of time's structure. We are used to thinking about time in different ways at once, holding several contradictory notions in our minds and applying whichever one is appropriate for an occasion. If we dissect these thoughts, however, the various distinct points of view can readily be seen.

For instance, consider the nature of holidays and other festive occasions. They simultaneously celebrate the periodicity of the world and its destructive and creative capacities. The rituals on these holidays, handed down from generation to generation, emphasize the point of view that the world is renewed year after year. On the other hand, these occasions are used to tally annual accomplishments and to celebrate increasing prosperity.

Finally, there is a sad awareness during holiday time that time is passing, that we are all growing older, and that death is nearing. During each festive occasion, all three points of view can be harbored at the same time without any sense of contradiction. Let's examine the subtlety of how this works by considering a concrete example.

Imagine a photographer who takes pictures at birthday parties. Suppose that he returns year after year to take pictures of the same family for the birthday celebrations of one of its children. He carefully compiles the photos for each party and places them in neatly arranged albums. Now suppose the photographer examines the pictures from two consecutive years. What he sees in the photos is a visual record of time. Depending upon the way he looks at the images, he can come up with several different ways of interpreting the events. These different approaches reflect the complexity of time.

For example, he may notice the *sameness* of the photographs. One year's set of photos may contain pictures of the same children, the same furniture, the same kind of cake, and the same decorations of the next year's set. It might turn out, for instance, that all the guests of all the parties are the same and the photos depict identical sets of people and things for different years.

Also, certain rituals represented in the photos presumably do not change from year to year. The photos may include several pictures of children cutting cakes, multiple snapshots of children singing, repeated images of identical party games, and so on. It thereby depicts the repetitive, cyclical nature of time.

On the other hand, the photographer can observe the aging process depicted in his photos. The new-born babies of one year's set of snapshots become the crawling toddlers of the next year's set. Older relatives appear more wrinkled or worn as time goes by. A careful study of the photographs reveals a host of irreversible changes of the family's appearance. These changes point to a unidirectional view of time: If a set of photos were scattered haphazardly over a table, without any markings of the years in which they were taken, their temporal order is still manifestly clear. There is only one logical historical arrangement of the photos: one in which the clearly younger versions of the relatives' images precede the clearly older ones. This points to the notion of time as destructive.

Finally, let's consider a third view that the photographer can take. The photographer's collection grows; each year new photos are added to it. Therefore, the photographer's profession represents a creative process in the sense that new images and new information are added to the world year after year. The story of the well-photographed family grows continuously as depicted in both the photos and the memories shared by its members. Thus, the passage of time can be seen as corresponding to increased complexity stemming from the acquisition of knowledge and experience. In this manner, time can be seen as creative.

The example of the photographer brings to light three ways of looking at time: as cyclical, as destructive, and as creative. These are three different "shapes" that time takes; they correspond to three different views of the world. As varied as that seems, however, the reality of time is far more complex. There are still other ways of depicting time.

For instance, consider the photos taken by the photographer in our story. Do they represent the reality of the instant in which the pictures were taken, or do they depict events of a previous period—an instant earlier? Initially, one might think that the events took place at the same time that the camera's shutter clicked. But because it takes a small but finite period of time for light to travel from the objects to the lens, the images captured by the camera actually represent time-delayed pictures. Thus, the idea of simultaneity, the notion that one can talk about two events taking place at the same time, is more complicated than one might think at first. Because of the finite amount of time that it takes information to travel, the idea of two events occurring simultaneously has a subjective component. This fact has a bearing on the nature of time.

Another question about time regards its apparent "speed," which seems to be affected by internal factors such as aging. Here one might consider, for instance, the time interval between the birthday parties as experienced by both the photographer and the children. One year for the children, full as it is of hundreds of new experiences and sensations, might seem like forever; the same year for the photographer might seem short and eventless. As people get older, years generally seem to roll by faster and faster because each one seems shorter and shorter. Clearly there are psychological and biological components to the experience of time which affect the human sense of duration. To understand time's nature, we must somehow find a relationship between subjective and objective descriptions.

Another issue regarding time relates to whether or not it is continuous. In light of our story, we ask the following question: Could the photographer record each instant of some event without leaving any gaps? Obviously not; no matter how quick the photographer would be, there must be instants between clicks of the shutter which he could not record. Suppose, however, that he could photograph each moment without exception. Would that form a continuous stream of images, or would there still be gaps? This relates to the broader question: Is time a uniform stream or is it formed of "atoms," distinct moments separated by imperceptible breaks? In other words, can time be subdivided again and again without stopping, or must there be a stopping point where the smallest unit of time will be reached? Thus the story of the photographer presents us with yet another puzzle regarding time: Is it continuous or discrete?

In this book, we shall look at the many forms of time and examine the contradictions of the conflicting views. Because the study of time is truly interdisciplinary, we shall compare and contrast the results of biological, geological, philosophical, psychological, historical, and theological inquiries. We shall draw from sources as diverse as the poetry of Milton, the holy writings of the Bible, the clinical notes of neurologist Oliver Sacks, the short stories of Borges and Wells, and the sketchbooks of Darwin. Because the focus of this book is the relationship of these ideas to contemporary themes in physics and cosmology, we shall above all emphasize how the varied temporal notions of modern physics mirror ancient debates over time's riddle. Finally, in the book's closing pages, we shall address the question of the relationship between one's approach to the question of time and one's attitude toward the issue of mortality.

Let us now examine the basic structure of the book. In the first chapter, the idea of circular time will be discussed. From the Indian notion of cosmic rebirth to the contemporary oscillatory universe model, the idea of temporal periodicity has a strong appeal. We shall trace its origin back to the fertility myths of early agricultural civilization.

Next we shall explore the linear picture of time. The ideas of irreversible change and historical development will be explored. We shall examine two approaches to the meaning of linear time: the pessimistic and optimistic visions. We shall trace the former approach back to the apocalyptic visions of the early Christian thinkers and the latter back to the progressive ideals of the Renaissance. We shall show how these temporal outlooks entered the world of science with the nineteenth century discoveries of the laws of entropy and evolution.

Then, in the fourth chapter, the ideas of causality and synchronicity will be presented. We shall examine how special relativity defines strict limits to the rate by which information can travel and how quantum mechanics may enable us to break those limits. In the fifth chapter, we shall examine the question of the flow of time. First we shall see how psychological time differs from clock time; then we shall explore the issue of time travel: the question of whether or not our conscious minds can journey freely to the future or past.

Finally, the multiforked and atomistic notions of time will be discussed. We shall consider both the idea that time is constantly subdividing into separate streams and the notion that time is composed of numerous particles. Each of these scenarios has its supporters among modern physicists.

In each case we shall compare certain philosophical and literary concepts with contemporary physical schemes. Thus, we shall show how the temporal themes of modern physics have roots in ancient images of the universe. It is truly remarkable how many parallels one can find between contemporary science and time-tested philosophy.

This book does not purport to be a comprehensive study of all of time's manifestations. A complete exposition of everything that has been written about time would take several lifetimes to compile and at least one lifetime to read. Rather, this work is a *survey* of the different ways of looking at time intended to draw out the similarities in the approaches of various fields of inquiry. Therefore, it is hoped that the reader will find the time to read more on this subject. The reader should also note that it is hard to "find" time.

Paul Halpern

Acknowledgments

I would like to acknowledge the generous support for this project given by the staffs of Hamilton College and the Philadelphia College of Pharmacy and Science. In particular, I wish to thank Mary and John De Forest, Peter Millet, Bernard Brunner, Allen Misher, Elizabeth Bressi-Stoppe, Doraiswamy Ratnasabapathi, Robert Cherry, Boris Briker, William Yarnall, and James Ring for their help and encouragement. I would also like to thank Max Dresden for introducing me to many interesting questions regarding cosmology and chaos, Roger Penrose for a fascinating and inspiring lecture at Syracuse University, and Jennifer Mitchell for her useful editorial suggestions.

Finally, I would like to acknowledge the encouragement and support of my family and friends including my parents, Stanley and Bernice Halpern, my brothers, Richard, Alan, and Kenneth, and my good friends, Michael Erlich, Fred Schuepfer, Donald Busky, Robert Clark, Kris Olson, Simone Zelitch, and Graham Collins, all of whom have made useful suggestions regarding this work.

Time is the One Essential Mystery.
JORGE LUIS BORGES

Time Journeys

1
The Circular Road

A History of Cyclic Time

A child emerges from the womb. It draws its first breath of air. This breath is followed by another and yet another. A small heart pumps blood through the child's tiny veins and arteries. The heart beats over and over again, forcing the blood around the body, bringing life to all of its parts. The rhythm of existence has been set into motion. Time has begun for the child.

At first, sleep comes to the child at odd moments, and the child's bodily patterns seem insensitive to the surroundings. Gradually, however, the pace of day and night set a strong rhythm for the child. Eventually sleeping at night and waking each day becomes an integral part of its life, so that even if it is removed from any exposure to the light of the sun, the infant continues to follow a 24-hour cycle of metabolism.

The repetitive motions associated with breathing, with the beating of the heart, and with daily sleep patterns are all examples of the clockwork rhythms called cyclic time. Cyclic or circular time is the periodic pace measured by a clock or a metronome. The time measured by a clock is fundamentally different from the time measured by a calendar or a history book. Cyclic time excludes the notions of progress and decay because it admits only a present, a present moment that is indistinguishable from any other moment. Any action that is repeated again and again without change sets the pace of a circular time. This pace may be as rapid as a heartbeat or as drawn out as the motion of the solar system around the center of the galaxy.

Clearly, many features of the world are cyclical. The earth rotates every 24 hours, causing a dramatic difference in light and heat. The spring thaw follows the winter frost every year. Plants wither away, only to bloom again in a few months. The tides advance and recede, drawn by the unrelenting pull of the sun and the moon. Birds journey south, then north, then south again in seasonal migration patterns.

Moreover, the rhythms of the heavens strongly influence the way we think and feel and the structure of our lives. All of us experience daily alterations in metabolism even when not exposed to environmental changes. Because we "know" on a deep level when we should eat and sleep, it feels strange to us if this pattern is disrupted.

The bodily mechanism which sets this rhythm is known as the circadian timing system. The word "circadian" comes from the Latin words *circa*, meaning "about," and *dies*, meaning "day." Circadian rhythms require about 24 hours per cycle and include patterns of sleep, wakefulness, food digestion, and energy metabolism. These patterns, although set by the solar rhythm, generally continue even in the absence of direct exposure to the sun.

Anyone who has taken a long intercontinental flight is keenly aware of what happens when the natural circadian rhythm is suddenly disturbed. The phenomenon of jet lag is, unfortunately, all too real for those who have traveled a distance of several time zones and, as a result, have had their internal clocks disrupted. The experience of lying awake at night when one, by the time indicated on the clock, should be sleeping is not too easily forgotten! Generally, those who must fly frequently try to follow regimens that help them ease their way into their new schedules. It becomes all too clear to them that they can't always take their circadian clocks for granted.

In addition to circadian processes, we are affected by a very wide range of biological clocks, such as women's monthly patterns of biochemical changes. In many ways, our internal life rhythms are deeply tied to the motion of our parent planet. Yet, unlike the manifestly eternal movements of our world, our internal mechanisms do slow down and eventually cease to function. Human mortality stands in stark contrast to the clockwork universe that surrounds us. We are born and we grow, age, wither, and die, never to be seen again. Time chews us up and spits us out, leaving only a carcass. We plunge headlong into an uncertain future, knowing that we can never return to our past. Our lives are characterized by constant change. As the poet Catullus wrote, "Suns may set and rise again. For us, when the short light has once set, remains to be slept the sleep of one unbroken night."

This duality between the timeless nature of the universe and the transient aspect of human existence has been a major theme of religious philosophy. Because the human experience has been one of looking beyond the apparent fragility of life to seek a transcendental permanent reality, the history of religion can be said to be a document of the struggle to understand and supersede mortality and unite with the everlasting truth of the cosmos. In other words, religion has sought to reconcile the fragility of life with the permanence of truth and "reality," the latter usually equated with the divine.

This search for the omnipresent is clearly seen in the fertility myths of the early agricultural societies. Primitive humans, leading brief and arduous lives, searched for immortal truths in the natural world around them. Noticing that, although human and animal lives were limited, plant life was perennial, they began to look for hidden meanings in agriculture. Feeling that plants possessed a kind of immortality, they developed medicinal potions based on fruits and herbs. Their goal was to claim this gift for themselves and thus to be able to live again and again in renewed forms of being. Because it was felt that the rhythm of vegetation furnished the key to the mystery of birth, death, and rebirth, fruits and herbs were thought to provide everlasting existence and knowledge of ultimate reality.

Other natural phenomena which experienced rebirth also played a strong role in primitive myths. It was noticed that the sun would die every evening and then would be reborn the following morning. That too was thought to be related to the elusive goals of immortality and cosmic awareness. The same held true for the motions of the moon and the stars; those celestial objects were often considered to be fortunate creatures that possessed the secret knowledge of how to evade death.

Since celestial and agricultural cycles played such a vital role in the religious ideas of early civilization, cyclical time, the time of the sun and the moon as well as that of the flowers and grains, was seen as holy time. All natural rhythmic processes, especially those believed to be connected with heavenly motion, were included in this worship of the everlasting circle.

Since seasonal rejuvenation was seen as an antidote to human decay, women, as bearers of life, were thought to have knowledge of this tonic. Women's menstrual cycles were considered to be links between the divine lunar rhythms and the earthly processes associated with fertility. Thus, female fertility was seen as being connected with the mystery of vegetation and the celestial motions which set the pace of cyclical time.

Considering this strong interest in the periodicity of agricultural life, it is no wonder that the harvest period played such an important role in Neolithic cultures. The new year festival, celebrated during the time of reaping, involved the idea of rebirth and the periodic renewal of the world including, in some societies, the myth of the return of the dead. Many of these traditions continue today in some cultures, although their original meanings are often distorted. It is interesting to note, for instance, that the Jewish New Year is still observed during the harvest season, coming just a few weeks before Succoth, the festival of gathering.

Periodicity played a vital role in other early religious traditions. In early Indian (Vedic) thought, annual sacrifices and daily fire offerings were integrally linked to the continuation of the world. The Brahmans felt that the immortality of the universe could be maintained only by priestly rituals at the fire altar.[1] As the noted religious scholar Mircea Eliade points out, it was felt that even "the sun would not rise if the priest, at dawn, did not offer the oblation of fire." In this manner a repetitive, ritualistic lifestyle became linked to cosmic regularity, which in turn was associated with immortality and perfect truth. Eliade refers to this concept as "sacred time."

Eliade points out that sacred time is cyclical and therefore lacks any temporal direction, unlike the one-way time corresponding to ordinary mortal existence. By engaging in repetitive actions, early humans could identify with the timelessness of the sacred realm. Eliade puts this in the following manner:[2]

> By its very nature sacred time is reversible in the sense that, properly speaking, it is a primordial mythical time made present. Every religious festival, any liturgical time, represents the reactualization of a sacred event that took place in a mythical past, "in the beginning." Religious festival implies emerging from ordinary temporal duration and reintegration of the mythical time reactualized by the festival itself. Hence, sacred time is indefinitely recoverable, indefinitely repeatable. From one point of view it could be said that it does not "pass," that it does not constitute an irreversible duration.

In other words, only by engaging in ritual could worshippers break free from the mortal sphere and picture themselves to be part of the immortal realm of holy truth. Because these ceremonies represented a link with the distant human past, as well as with an imagined timeless sacred prehistory, they were considered to be ways of transcending death itself: Sacred reversible circular time could replace profane irreversible linear time.

As Hindu philosophy developed, this preoccupation with daily and seasonal rhythms became incorporated into the notion of the wheel of samsara: the periodicity of natural being. In this Indian tradition, everything exhibits birth, life, death, and rebirth. A human lifetime is considered to be an immeasurably small part of an endless chain of reincarnation.

Human history and the changing patterns of the cosmos were also seen as circular. It is a small step to proceed from the belief in the sacredness of the seasonal death and rebirth of nature to a belief in human rebirth and finally to a belief in societal rebirth. According to the Hindu writings, every 4,320,000,000 years, a period known as a *kalpa* (world cycle), the universe is destroyed and recreated. Hence, in this view, all of human history represents but a small and insignificant part of cyclical eternity.

These beliefs are shared by other ancient peoples that have occupied the Indian subcontinent. In the philosophy of the Jains, a non-Hindu Indian culture, the world cycle is pictured as a serpent (*sarpin*) devouring its own tail. For the Jains, as well as for the Hindus, this ceaseless circular flow of the cosmos goes on forever. That can be contrasted with the Buddhist world view, in which this cycle of creation and destruction is seen as continuous. Buddhism is a religion based upon the sacredness of contemplation. By meditating, one realizes the ever-changing nature of one's thoughts, which in turn leads one to contemplate the impermanence of all things. For the Buddhist the continuity of human life is an illusion, whereas the self is a succession of ideas and experiences. One can obtain inner peace only by letting go of the illusion of continuity and accepting the ceaseless pace of change.

The structure of Buddhist meditation clearly reflects that point of view. Typically, during the meditation practice, one focuses on the flow of one's breath, sensing its ceaseless rhythmic nature. Any thoughts that arise during the meditation are seen as temporary deviations from an eternal cyclic process. The breath sets the pace of a new perception of the continuous destruction and creation of the natural world.

In other forms of meditative practice, the repetition of a word called a mantra achieves the same purpose. By chanting (or thinking) a meaningless sound again and again, one hopes to break free of the illusory chain of ordinary thought (leading certainly to death) and to enter the realm of infinite repetition (leading to eternal existence).

This hope is reflected in the symbolic function of the Sanskrit mantra "om," which is formed by the combination of an open sound "oh" and a closed sound "mmmm." In opening one's mouth for the oh, one is creating (giving birth to) a sound; in closing one's mouth again for the mmmm, one destroys the sound. Hence, chanting this mantra represents the act of continuous creation and destruction.

In its adherence to a belief in the sacredness of repetition, Buddhism, Hinduism, and Jainism shared a similar belief in cyclic time. In fact, cyclic time, with its strong links to nature and to immortality, had an enormous appeal to almost every civilization in the ancient world; it represented the common hope of ancient peoples to connect their destiny to the eternal. Even in ancient Greece, where there were many divergent and conflicting views of temporal progression, the notion of periodic time exerted a strong pull.

The earliest records of Greek philosophy date back to the sixth century before Christ, when the Greek civilization extended throughout the Aegean islands, Asia Minor, and southern Italy as well as the Greek peninsula. The Greeks were very much interested in the patterns of nature and the relationship between the natural and human realms. Much of early Greek cosmology was derived from equating human biological development with the evolution of the universe. As human life reproduces itself, so events in the universe were seen as links in an eternal chain of generational procreation.

Little is known of the life of Heraclitus of Ephesus, one of the earliest Greek philosophers. Heraclitus lived at approximately 500 B.C. during the time of the Persian conquest of Asia Minor. As a Greek in a city conquered by the Persians, he was exposed to both the Greek myths and Zoroastrian traditions. Borrowing the Zoroastrian notion that the universe is a struggle between two forces (good and evil), Heraclitus postulated that the universe is in a constant state of conflict between opposing principles. The world is constantly changing and yet eventually must return to the same state. Heraclitus believed that, in spite of continual chaos, there is a cyclic order in all things. This, he felt, is because things tend toward their opposites. When the opposite of an opposite is reached, then events repeat themselves. As Heraclitus said:[3] "Cold things grow hot, hot things grow cold, the wet dries, the parched is moistened," but "the way up and the way down are one and the same." Thus, in the Heraclitean view, everything must return to its original condition in a repetitive manner.

About fifty years after the time of Heraclitus, the Sicilian philosopher Empedocles appeared on the scene. Empedocles, like Heraclitus, believed in a cyclic time driven by opposites. For Empedocles, these opposites were love and strife. Love brings separate elements together; strife divides united bodies. In the cosmology of Empedocles, the earth passes through various eras in which one or the other of these principles dominates. In a sort of metamorphosis that occurs, the human race appears via a series of evolutionary steps, emerging from species of half-formed humans and animals that separate and recombine during these eras. These intermediate monstrosities are formed by inadequate combinations of love and strife. Finally, in the present era, humanity as we know it emerges. However, it is not expected that the current period will last forever, since this evolution is thought to be eternal, with new stages following as the process continues. Therefore, according to Empedocles, there is no beginning or end to the world, so time is circular.

The post-Socratic Greek philosophers continued the tradition of viewing time as cyclical. Plato, in his *Timaeus*, linked time to the succession of natural events, and in particular to planetary motion. He thought of earth and the heavens as imperfect reflections of divine order: The movement of the planets comes closest to being an image of this reflection. Plato viewed time as inseparable from that periodic motion. The progression of time as expressed by planetary orbits is, in his words, "the moving image of eternity."[4] It is for that reason he believed in a recurrent universe in which the world is periodically renewed. Once again, we see how, in the early civilized world, circular time was equated with the divine.

The destructive power of water played an important part in Plato's belief in recurrence. It was a common belief, in the time of Plato, that floods periodically destroyed the earth. In his *Timaeus*, Plato cited records of the partial destruction of the world by flooding. Plato, and later Aristotle, argued that this was evidence for periodic world devastation.[5]

Unlike Plato, who thought of nature as a mere reflection of eternal truth, Aristotle believed in the physical reality of the natural realm. Nevertheless, his view of time bears a strong similarity to Plato's. Time, for Aristotle, is a measure of change. Since every change requires a cause and that causal action requires yet another cause, time has to be without beginning or end. Because a circle is an object without endpoints, Aristotle argued that time must be related to circular motion. Because only the planets and stars describe perfect circles in their motion, he thought they quantify time. Aristotle referred to change as "generation and corruption," and he wrote that "the continuity and eternity of generation and corruption is an imitation of the continual circular motion of the eternal heavenly bodies."[6] Aristotle's powerful arguments served as a definition of temporal reality for hundreds of years.

Greek culture reflected those views of time. Since time is cyclic and perfect, it was argued that a political system should be as unchanging as possible and that there should be no need for historical progress or revolutionary growth. Therefore, most Greek philosophers, including notables such as Pythagoras and Heraclitus, advocated a cyclic concept of human history.

The Greek romances, which appeared hundreds of years after the time of Plato and Aristotle, also displayed a preoccupation with timelessness. As the literary theorist Bakhtin points out,

> All of the action in a Greek romance, all of the events and adventures that fill it, constitute time-sequences that are neither historical, biographical, nor even biological or maturational.... In this kind of time, nothing changes: the world remains as it was, the biographical life of the heroes does not change, their feelings do not change, people do not even age. This empty time leaves no traces anywhere, no indications of its passing.[7]

In other words, Greek romantic heroes were much like present-day comic book superheroes. In the same way that contemporary comic book aficionados would be stunned if Superman radically altered in appearance, aged, or died, Greek audiences expected their fictional heroes to never change. They relished the idea that these heroes belonged to an imaginary timeless world free from the tyranny of the clock. Thus, Greek romances continued the classical tradition that the sacred realm exists outside the constraints of ordinary linear time.

Eternal Return

Circular time has been both compelling and frightening to philosophers, theologians, writers, and scientists throughout all history. A number of nineteenth- and twentieth-century philosophers and writers found the idea of eternal repetition to be of great interest. One of the most prominent advocates of circular time was the German philosopher Friedrich Nietzsche. Nietzsche, who led a troubled life, was one day struck with horror at the thought that he might be compelled to face his miseries again and again. Because time is infinite but the world is finite, all the events of the world must repeat themselves an infinite number of times, it seemed to him. He spent much of his life trying to cope with this sudden awareness.

Referring to his concept as the idea of *eternal return*, Nietzsche believed that he originated the notion, in spite of the long history of its development. It was, however, clear to his contemporaries that Nietzsche borrowed heavily from Indian and Greek traditions of cyclic time. In particular, eternal return resembles the Heraclitean belief in a periodically repeating ending of the world out of the all-consuming world conflagration.[8]

In a similar manner, Nietzsche thought history to be a series of constructive and destructive epochs. During the constructive periods, the greatest achievements of humanity manifest themselves because those with the greatest "will to power" emerge and build an increasingly complex civilization. After the era of construction reaches a high point, forces of destruction tear down what was created and the cycle begins anew.

It was felt by Nietzsche that the fragility of all of the greatest human creations mandated their eventual collapse. That is the reason, he argued, why the so-called eras of construction are temporary. However, he said, in spite of their instability, the achievements during these productive periods are both real and important.

Newtonian physics provided Nietzsche with a justification for his ideas of eternal recurrence; namely, the notion that energy cannot be created or destroyed. Nietzsche felt that "the law of the conservation of energy demands eternal recurrence" because the energy of the world must continuously recycle itself. Thus, as Nietzsche would point out, nothing is new under the sun. Everything that exists in the world has already been here in some other form.

Advancing another argument for a cyclical world, Nietzsche felt there to be only a finite number of possible events in the universe. If the universe consisted of a finite number of particles, he supposed the set of those particles would eventually exhaust all possible positions and interactions. By necessity, the particles would have to repeat their early behavior an infinite number of times, just as a pendulum constantly repeats the same motion. Nietzsche expressed this in his "will to power."

> If the world may be thought of as a certain definite quantity of force and as a certain definite number of centers of force...it follows that it must pass through a calculable number of combinations. In infinite time, every possible combination would at some time or another be realized; moreover it would be realized an infinite number of times...A circular movement of absolutely identical series is thus demonstrated: the world as a circular movement that has already repeated itself infinitely often and plays its game ad infinitum.[9]

Thus Nietzsche imagined the world to be a sort of chess game. Even though millions of possible moves and scenarios, more than any player could possibly see in his or her lifetime, are possible in chess, the possibilities are finite and limited in number. If one were to play chess trillions and trillions of times, eventually the games would be repeated again and again. The millionth game, for instance, might be identical with the fifth game. Therefore, it would appear that time is cyclical because the events of the chess game would be duplicated periodically.

The scenario was somewhat anxiety-provoking for Nietzsche, who imagined living his life, with all of its pain, an infinite number of times. That greatly troubled him until he realized that eternal return provided a type of immortality. By existing again and again, one has a permanent place in the universe, a place that would be denied if one were to live life only once. Here Nietzsche examines the anxiety that the idea of eternal return brings and ponders a solution:

> What if some day or night a demon were to steal after you into your loneliest loneliness and say to you: "This life as you now live it and have lived it, you will have to live once more and innumerable times more; and there will be nothing new in it, but every pain and every joy and every thought and sigh and everything unutterably small or great in your life will have to return to you…"[10]
>   Would you not throw yourself down and gnash your teeth and curse the demon who spoke thus?…Or how well disposed would you have to become to yourself and to life to crave nothing more fervently than this ultimate eternal confirmation and seal?

Two twentieth-century authors of fiction, the Argentine writer Jorge Luis Borges and the Czech novelist Milan Kundera, have been fascinated by the idea of eternal recurrence and have based some of their writings on the works of Nietzsche. Both Borges and Kundera find that the idea that time repeats itself leads to difficult issues regarding the human condition.

In Kundera's novel *The Unbearable Lightness of Being* it is suggested that the principle of eternal return leads to a burden of unbearable responsibility. "If every second of our lives recurs an infinite number of times, we're nailed to eternity as Jesus Christ was nailed to the cross," Kundera writes.[11] On the other hand, without eternal repetition, our lives would be ethereal because they would occur just once. In that case, we would lead our lives knowing that our actions were doomed to be forgotten. Kundera finds horror associated with both possibilities.

In light of the fact that all works of art would be predestined, Borges, on the other hand, argues that cyclical time implies that human creativity is meaningless. Given an infinite amount of time, all possibilities, including the writing of great and "original" novels, must occur. This theme is suggested in several of his short stories. In "The Immortal" he writes, "Homer composed the Odyssey; if we postulate an infinite period of time, with infinite circumstances and changes, the impossible thing is not to compose the Odyssey at least once."[12] Thus every creative work is merely a fortunate meaningful coincidence brought about by pure chance.

In the short story "Pierre Menard, Author of Quixote," Borges imagines an author to rewrite Cervantes' work *Don Quixote* word for word. Menard manages to compose *Don Quixote* through his own creativity, not merely to transcribe the novel. Once again, Borges emphasizes that creativity is a meaningless concept in light of the repetition of history.

Circular time can thus be thought of as a set of funhouse mirrors, reflecting our likenesses back and forth an infinite number of times. Gazing into these mirrors, we see ourselves again and again in images stretching off into a boundless panorama of our reflections. Thus, each of our actions, reflected in this looking glass world, seem to be repeated over and over for all eternity. That is why funhouse mirrors are the source of so much anxiety and confusion and, of course, fun.

We can imagine Nietzsche stepping into the mirror hall of such a funhouse. At first he would be terrified as he saw his troubled gaze reflected again and again in his surroundings. The miserable looks on the faces of his specters would make a mockery of his own pain and misfortune. Eventually, though, Nietzsche would stand tall before his audience. He would turn and salute all of his comrades of the struggle as he realized that power lies in numbers. In unison, Nietzsche and his reflections would find meaning in their synchronous mutual actions; because they appeared to be taking place over all of space, they would seem to be of profound importance.

Kundera, on the other hand, would take another attitude toward being in such a mirrored room. He would feel at once that his privacy was violated because everything that he would do would appear before him an infinite number of times. In an ordinary chamber, he could afford to act foolish and engage in harmless minor transgressions without carrying the burden of conscience. In the hall of mirrors, however, all his acts would be multiplied by the crystalline surfaces only to be scrutinized by the myriad of intrigued gazes. Thus, the comforting solitude of his actions would be lost as he stood amid the kaleidoscope of images.

Finally, Borges, on entering the funhouse, would suffer a severe identity crisis. At first he would wonder if he were the real Borges or if the reflections might be equally real. Then he would begin, pen in hand, to jot down his thoughts regarding his confusing predicament. Finding all of his images beginning to write down their thoughts as well, he would conclude that even the creative process is meaningless if it is infinitely repeated. In the end, one of the mirror images of Borges would stand up, turn around and leave, feeling quite puzzled by these events!

If we substitute temporal repetition for spatial repetition, we can see why all three of these influential writers, Borges, Kundera, and Nietzsche, found the possibility of eternal recurrence to be disturbing. Clearly then, each of them took this idea somewhat seriously and believed that there might be some possibility that the events of the universe would happen again and again. That is a reflection of the fact that the idea of circular time has had a strong influence on our way of thinking.

This influence stems from a number of factors. First, as we have seen, cyclical time is linked to natural clocks and celestial patterns, all of which continue to amaze us with their regularity. Second, we have all been influenced by the philosophy of our ancestors, many of whom felt that cosmic and human history repeat themselves. Finally, for nineteenth- and twentieth-century humanity, we must reckon the influence of the writings of science, particularly the highly esteemed works of Sir Isaac Newton and Professor Albert Einstein. Both physicists argued that the motions of the universe can be explained by simple, reversible, and unchanging laws and that the universe must therefore be repetitive and predictable in its behavior.

Time and Classical Physics

Considering the influence of the temporal notions expressed in Greek culture and the prominence of Aristotle's cosmological theories, as well as the elegance and simplicity of the primitive notions of cyclic time upon which they were based, it is no wonder that the formulators of the laws of classical mechanics, including Isaac Newton, drew a clockwork portrait of the cosmos. The ideal equations of motion were considered to be independent of time, displaying time reversibility. That is, the universe should look the same if viewed either backward or forward in time, just as the recorded ticking of a clock would sound the same played either backward or forward. All of the laws governing the universe would display this feature.

That is not to say that the work of Newton, the father of classical mechanics, was merely an extension of Aristotle's concepts. Newton's ideas about time vastly altered future considerations of time's properties. He postulated that there exists an absolute time that is independent of the motions of any celestial bodies. Newton divorced this notion of time from the common time of the apparent stellar procession and the mundane time of an ordinary clock. According to Newton, absolute time is a fixed standard by which all activities in the universe can be measured.

By severing the ties between time and celestial motion, Newton demystified the behavior of the heavenly bodies. Instead of considering this behavior to be self-evident in its simplicity, Newton questioned its underlying mechanisms. In doing so, he found simple mathematical relations which elegantly produce the cyclical motions of the planets as their direct consequence. These famous relations are known as Newton's laws of motion and Newton's law of universal gravitational attraction. Newton's laws of motion are as follows:

1. Every body continues in its state of rest or of uniform motion in a straight line unless it is compelled by a force to change that state.

2. Change of motion is proportional to the applied force and takes place in the direction of the force.

3. To every action there is an equal and opposite reaction or the mutual actions of two bodies are always equal and oppositely directed.

Newton's law of gravitational attraction states that objects exhibit a gravitational force toward each other in inverse-square proportion to the distance between them. Together with the laws of motion, this dictates that the planets must travel in ellipses around the sun and that the moon must revolve around the earth. Thus, even though cyclical time is not a supposition of Newton, cyclical motion is a consequence: Central, attractive forces, such as gravity, guarantee repetitive behavior for celestial objects.

Newton's laws exhibit complete time reversibility. To understand why that is true, picture the motion of a ball on a billiard table. If one hits the ball so that it bounces off a cushion, it will move in a certain direction with a certain speed. If one could then reverse the direction of the ball at the same speed, the ball would precisely trace its path backward to its starting position. All of the forces on the ball would be the same. One could not distinguish the backward motion from the original motion, because the ball would look as if it were replaying its original motion backward in time. Its motion is geometrical and therefore symmetric in time.

Solving for the motion of a frictionless pendulum is a common problem in Newtonian classical mechanics. If a pendulum starts at a certain angle from the vertical and is released, one can calculate its speed and position for all future times. In this ideal case, the motion is totally determined; that is, one can know the future completely by knowing the present. A principle that guides the path that the pendulum takes is that of the conservation of energy, which is a direct consequence of Newton's laws of motion. One can show that energy can be neither created nor destroyed; it can only be transformed from one type to another. Here the two types of energy are the energy of motion and the energy of position of the pendulum. This ensures that the pendulum undergoes the same behavior again and again, because otherwise energy would be gained or lost.

That can be pictured as an hourglass in which the upper part of the hourglass is labeled "energy of position" and the bottom part is labeled "energy of motion." The upper part of the hourglass is filled with sand (energy of position), which corresponds to the fact that the pendulum is held at a certain angle from the vertical. The bottom of the hourglass starts out empty, because the pendulum is not moving and there is no energy of motion. When the pendulum swings, its energy of position is transformed into energy of motion. The sand in the hourglass flows from top to bottom until the top is empty and the bottom is full. Now all of the pendulum's energy is energy of motion, because the pendulum is at its lowest position and is moving its fastest.

When the pendulum swings back up to the other side, it slows down. Its energy of motion is now being converted back into energy of position as it rises. Similarly, the hourglass can be turned over, causing the sand in the energy-of-motion half to flow once more into the energy-of-position half. Just as the sand from the hourglass can flow back and forth from one half to the other, so the pendulum can keep on swinging, constantly transforming its energy from one form to another. Thus, conservation of energy guarantees temporal periodicity.

There is a clear relationship between the regularity of time and space in the Newtonian picture. Newtonian space is uniform in character. Space is populated with stars, planets, and other objects fairly evenly distributed. Thus, one cannot distinguish between one part of space and another. Since energy is conserved, these celestial objects exhibit essentially periodic motion. Therefore, Newtonian time also is uniform. It is impossible to tell the difference between one point in time and any other, because the Newtonian universe looks pretty much the same for all times. There is no fixed yardstick for Newtonian space or for Newtonian time.

This similarity between space and time led the eighteenth-century mathematician Lagrange to regard time as a fourth dimension of space, thus reducing the role of time in physical theory. Time and space would be on an equal footing, with temporal displacement assuming a spatial, geometric character. Other physicists followed suit in developing time-independent physical laws. By the end of the 1800s, there was an increasingly strong tendency to attempt to write all physical equations in a time-invariant, geometric manner.[13]

The Union of Space and Time

By the turn of the twentieth century, the notion of linking time and space as one entity had become increasingly attractive. In the classic novel *The Time Machine*, by H. G. Wells, the protagonist describes a four-dimensional geometric amalgamation of space and time. "There is no difference between time and any of the three dimensions of space except that our consciousness moves along it," he remarks.[14] In just a few decades this fantasy of Wells became scientific theory. Utilizing the geometric motions of Lobachevsky and Riemann, Einstein developed a fusion of time and space, called space-time, and incorporated it into his theories of special and general relativity.

In Einstein's theory, any two events in the universe have a four-component distance: three spatial components and one temporal component. The spatial components are what we might call length, width, and height; the temporal component is the duration. One could draw a four-dimensional map of the universe by plotting the distances between any two events. This detailed description of all distances is called a *metric*. For example, the distance between the Battle of Hastings and the Battle of New Orleans would be about 5000 miles west, 2000 miles south, a difference in elevation of less than a mile, and a time difference of 8 centuries.

One could theoretically construct a model of the universe for all times. This model would be static within its four-dimensional realm. In such a manner, the concept of time as a flow would be wholly eliminated. The fact that we perceive time as having a direction would be seen as a human frailty, a complete illusion. As the hero of *The Time Machine* puts it:

> Clearly any real body must have extension in four directions: it must have Length, Breadth, Thickness, and Duration, but through a natural infirmity of the flesh we incline to overlook this fact. There are really four dimensions, three which we call the three planes of Space, and a fourth, Time. There is, however, a tendency to draw an unreal distinction between the former three dimensions and the latter, because it happens that our consciousness moves intermittently in one direction along the latter from the beginning to the end of our lives.

Wells, the incorrigible forecaster of human destiny, had accurately predicted one of the major premises of general relativity, 20 years in advance!

We should point out, however, one nontrivial mathematical difference between space and time in Einstein's theory. In ordinary space, when one calculates the distance between two points, one can use the so-called Pythagorean theorem: the square of the total distance is the sum of the length squared plus the width squared plus the height squared. In space-time, the square of the total metric distance is the sum of the length squared plus the width squared plus the height squared *minus* the duration squared. Thus, time carries a minus sign in the metric and is not exactly on an equal footing with space. Nevertheless, it is not clear how this mathematical distinction in general relativity affects whether or not time seems to flow. Therefore, time can be said to have a geometric, spatial character in this theory.[15]

Einstein's theory of general relativity presented a new way of looking at gravitation. In Newton's theory of gravitation, every entity which has mass exerts a force on every other object (Newton's law of universal gravitational attraction). Then each object changes its motion according to the resultant of all of the forces acting on it (Newton's second law of motion). If there are no forces on it, an object moves in a straight line with constant speed, because that is the shortest path through space (Newton's first law of motion). In Einstein's theory, on the other hand, every entity which has mass distorts the fabric of space-time. All objects pursue the shortest distance through space-time. Since massive objects warp space-time, the shortest path through space-time is not always the shortest path through conventional uncurved space.

One can envision this by considering an analogy. Imagine space-time to be a large trampoline. Two children are playing marbles on the trampoline. The goal of their game is to shoot the marbles as straight as possible, which can be easily done on the flat elastic surface. Each marble takes the shortest path along the surface, a straight line. Suddenly, a large rock falls on top of the trampoline, distorting its surface considerably. If the children attempt to resume their game, they find that the marbles no longer travel in straight paths along the altered surface; they curve in the direction of the unwanted intruder! By analogy, we see why massive objects distort space-time and cause other objects to move in curved trajectories.

Einstein's theory has been found experimentally to be more accurate than Newton's theory. However, the qualitative picture of planetary motion that arises from Einstein's theory is essentially the same as that from Newton's scheme. One would expect that to be the case because of the remarkable predictive success of Newton.

Let's compare how Newton and Einstein would each explain the motion of the earth around the sun. In Newton's description, the sun is seen as a generator of gravitational force because it has mass. The earth, wanting to continue in straight-line motion, finds it cannot, because of the sun's pull. Therefore, its motion changes, and it is forced to travel in an ellipse.

In Einstein's scheme, the sun is viewed as a distorter of the space-time fabric of a region of the universe. That means the sun alters the metric of the region. The metric describes all of the geodesics, or shortest paths, in the area. In a flat space-time, the earth travels in a conventional straight line in space, which is the shortest distance. However, because the space-time around the sun is distorted, the shortest distance in space-time becomes an ellipse. That is the path the earth is forced to take. The sun, in having mass, changes the shortest distances between all of the objects in the area. The concept of force, as a pull between two objects, plays no part in Einstein's gravitational theory. To sum this all up, we quote the classic textbook on relativity, written by C. Misner, K. Thorne, and J. Wheeler, "Space tells matter how to move....Matter tells space how to curve."[16]

Mathematically, the delineation of Einstein's theory is a bit subtle. It involves precise relationships among several mathematical objects representing the distribution of all of the mass and energy in the universe and the distances between any two events in the cosmos. Indirectly, the distribution of matter and energy in the universe creates the space-time web which dictates the motion of all celestial objects. This procedure is very difficult to follow exactly, except in certain simple cases. One common calculation involves the structure of space-time around a massive object such as the sun or what is known as a black hole. From that, the motion of other objects around this first object can be deduced. Another case is the determination of the behavior of the universe as a whole, a subject known as cosmology.

Time and the Cosmos

In 1917, Einstein attempted to apply the theory of relativity to the cosmological question of the universe's behavior. He examined the spatial evolution of the universe with respect to cosmic time, and he found what he sought: a stable solution to his equations which maintained the same character for all eternity. Based upon Einstein's paper, the completely clockwork, static universe dreamed of by the ancients was finally realized mathematically.

Unfortunately for Einstein and for those who savor a static world order, the paper contained a major error! In 1922, Alexander Friedmann corrected the mistake and found that Einstein's static universe wasn't stable, that it would expand or contract, given the slightest perturbation. Einstein quickly took steps to make amends by adding a new term to his equations. The term, called the cosmological constant, stabilized the universe, making it static. It balanced gravitational attraction with cosmic repulsion.

The addition of the cosmological constant term proved quite controversial. Many cosmologists felt that there was no physical motivation for the existence of an extra repulsive term in the Einstein equations. Einstein himself came to admit that the introduction of the term was a grave error. A series of astonishing astronomical discoveries dealt a devastating blow to the idea of a static universe, causing the Einstein model to rapidly lose favor.[17]

During the 1920s, astronomers began to realize that the universe is composed of galaxies: large, distant groupings of stars and gases. Between 1912 and 1925, V. M. Slipher measured the Doppler shifts in the spectra of more than twenty galaxies. (Doppler shifts are the changes in spectral lines when an object approaches or moves away at a certain velocity. This shift occurs in sound waves as well as light waves. It is for this reason that a train whistle changes in pitch depending upon whether the train is coming toward you or moving away from you.) To Slipher's surprise, all the galaxies were found to have red-shifted spectra. In other words, all the galaxies that he cataloged were found to be moving away from our galaxy. The astronomer E. Hubble examined more galaxies and found that all, except the closest, are receding from us. Furthermore, Hubble found the rate of recession to be proportional to the distance from our galaxy.

There is one obvious explanation for this phenomenon. Because our galaxy does not occupy a special place in the cosmos, all galaxies must be moving apart, as if from an explosion. If one were to trace the explosion to its origin, one would find that all the galaxies were at one time much closer together than they are now. In the Big Bang theory proposed by Gamow and Alpher in 1948 as a modification of an earlier idea by Lemaitre, the universe began in a state of extreme density. In the original proposal, this state was described as a hot fireball of subatomic particles called neutrons. An explosion took place, hurling all of the matter and energy of the universe outward. Finally, somehow, this matter coalesced into galaxies, stars, and planets.

As one might expect, many scientists and philosophers found the new "proof" that the universe had a beginning to be most unsettling. If every event in the world had a cause, what was the cause of the Big Bang? How could all of the matter and energy in the universe suddenly appear out of nowhere? Didn't that violate the law of conservation of matter and energy?

The astronomers Hermann Bondi, Thomas Gold, and Fred Hoyle provided an answer to those questions.[18] They considered the acceptance of the spontaneous creation of the universe to be a sudden rejection of the vast storehouse of physical intuition that had accumulated since the work of Newton. In particular, they felt acceptance of the Big Bang theory would negate the notion that the physical laws that exist today existed in the same form in the past. If the universe were constantly evolving, that could not be guaranteed. Therefore, they postulated, it would be more aesthetically pleasing to assume that the universe has always existed in basically the same form with the same physical laws and the same size.

The steady-state cosmology that they proposed has neither beginning nor end. It satisfies an axiom constructed by Bondi and Gold: the perfect cosmological principle. This principle states that the universe appears, on the average, to be the same at any given time. Since the perfect cosmological principle mandates that the universe never changes, it provides a means of restoring the clockwork model of Newton while taking into account the red shift caused by the recession of the galaxies.

In order to explain the red shift, Bondi and Gold introduced the idea that matter is continuously created in small amounts everywhere in the cosmos. This matter joins with existing gases in the universe to form new galaxies which recede from all of the other galaxies as the universe expands. However, since new galaxies are always being created, the average number of galaxies in any given region stays the same. Thus, the universe maintains the same character for all time.

The created matter needed for this process is so small that it is beyond detection. Therefore, the law of conservation of mass and energy would be violated by this theory. However, as Bondi and Gold argued, the Big Bang theory violates those laws in a major way at the origin of the universe. It is more logical, they thought, that the laws, if violated, would be violated in small ways constantly, rather than in a large way only once.

In order to address the question of conservation of mass, Hoyle went a step beyond Bondi and Gold. He developed a modified version of the steady-state model which did not violate any conservation laws. To do so, he modified Einstein's equations by adding a term. The added term allows for the continual creation of matter by drawing on another source of energy and thereby preserving the overall conservation of matter and energy.

For years, the steady-state theory offered a somewhat credible alternative to the evolutionary universe for those who found the notion of time having a beginning somewhat hard to swallow. Most recently, though, the theory has lost credibility because many of its predictions have turned out to be false. There are, however, other credible models which similarly fulfill the goal of an eternal cosmology.

One model that has remained viable is that of a cyclical universe. Here the universe expands from a point, reaches its full extent, and then recontracts to a point. It then expands again; the process occurs ad infinitum. Many solutions of the Einstein equations fit that prescription. We can explain the current expansion of the universe by assuming that we are in an early phase of a cycle. This theory, also known as the oscillating universe idea, was originated by Alexander Friedmann in 1922. Only a universe with sufficient density can exhibit such behavior. It is not known whether or not our universe has enough matter and energy to cause a collapse.

An interesting question with regard to an oscillating universe is this: What happens to the passage of time during the recollapse? Does time continue to pass, or does it reverse itself? If the latter is the case, will we live our lives again in reverse order during the second part of the cycle? This possibility has been suggested by Thomas Gold. Stephen Hawking, one of the most brilliant cosmologists of the century, has been, until recently, an advocate of that view. On the other hand, if time continues to pass in the normal manner during the collapse and reexpansion, will the universe, in the next cycle, possess the same physical laws? John Wheeler advocates the viewpoint that it may not. According to him, the universe is "reprocessed" as it bounces, so that it emerges with new laws and new physical constants. Each expansion and reconstruction produces a new set of conditions. Some universes, because of hostile initial factors, may not even allow life to form or permit planets to be created.

It seems that our discussion of cyclic time has come full circle! The contemporary cyclical universe theory is remarkably similar to the Indian concept of yugas or world cycles, the Greek notion of cyclic time expressed by philosophers such as Plato and Heraclitus, and the eternal return of Nietzsche. In each case the universe proceeds through an infinite sequence of creative and destructive epochs. Change is possible, but in the end everything returns to the same state via a catastrophe.

In seeking immortality, humanity has looked to the heavens. In our rituals we have endeavored to capture the rhythm of the stars for ourselves. We have sought eternal truths untarnished by the currents that erode our corporeal selves. Our physics has mirrored this quest for timelessness. Attempting to capture the celestial motions on paper, we have developed physical equations that are independent of temporal progression. In our search for models of the universe, we have been attracted to those without beginning or end.

Yet our desire for eternity is matched by our horror at the thought of an eternal recurrence in which our lives and our works are manifestly futile. If the universe does repeat itself again and again, all that we create will return to dust someday. One possible alternative to the despair of nihilism is the hope provided by religious faith. Let us now consider how some of the western religions have viewed the meaning of time and history.

References

1. Ruth Reyna, "Metaphysics of Time in Indian Philosophy," in Jiri Zeman (ed.), *Time in Science and Philosophy*, Elsevier Publishers, New York, 1971.

2. Mircea Eliade, *A History of Religious Ideas*, University of Chicago Press, Chicago, 1978.

3. James B. Wilbur, *The Worlds of the Early Greek Philosophers,* Prometheus Books, Buffalo, N.Y., 1979.
4. F. Cornford, *Plato's Cosmology,* Harcourt, Brace & Co., New York, 1931.
5. A. E. Taylor, *A Commentary on Plato's Timaeus,* Oxford University Press, Oxford, 1928.
6. Aristotle, C. J. Williams (trans.), *De Generatione et Corruptione,* Clarendon Press, Oxford, 1982.
7. M. M. Bakhtin, *The Dialogic Imagination,* University of Texas Press, Austin, 1981.
8. F. Copleston, *Friedrich Nietzsche, Philosopher of Culture,* Harper and Row, New York, 1975.
9. Friedrich Nietzsche, A. Ludovici (trans.), *The Will to Power,* T. N. Foulis, London, 1910.
10. Richard Schacht, *Nietzsche,* Routledge and Kegan, London, 1983.
11. Milan Kundera, *The Unbearable Lightness of Being,* Harper and Row, New York, 1984.
12. Jorge Luis Borges, "Library of Babel," "Pierre Menard, Author of Quixote," and "The Immortal," in *Labyrinths.* Copyright © 1962, 1968 by New Directions Publishing Corporation, New York. Reprinted by permission of New Directions Publishing Corporation.
13. G. J. Whitrow, *The Natural Philosophy of Time,* Harper and Row, New York, 1963.
14. H. G. Wells, "The Time Machine," in *Three Prophetic Novels,* selected by E. F. Bleiber, Dover, New York, 1960.
15. Geza Szamosi, *The Twin Dimensions,* McGraw-Hill, New York, 1986.
16. C. Misner, K. Thorne, J. Wheeler, *Gravitation,* Freeman, San Francisco, 1973.
17. Jacques Merleau, *The Rebirth of Cosmology,* Alfred A. Knopf, New York, 1976.
18. Jayant Narlikar, *Introduction to Cosmology,* Jones and Barlett, Boston, 1983.

2
The Road Slopes Downward

The Demise of Circular Time

According to the Jewish, Christian, and Moslem religions in their most traditional form, the world was created from nothing about 5800 years ago. On the sixth day of creation, man was formed from the dust of the earth and woman was created from man. Placed in the Garden of Eden, they lived an idyllic existence free from suffering and disease. In their innocence, they lacked full self-awareness and any sense of change. Nothing in their lives could mark the flow of time. Without birth or decay, every day was like every other day.

It is not clearly stated in the Bible whether or not Adam and Eve were truly immortal or were simply unaware of death. In the latter case, they would be in the same position as most living creatures. For it is true that, as Borges has said, "To be immortal is commonplace; except for Man, all creatures are immortal, for they are ignorant of death."[1]

Adam and Eve could not eat the fruit of two trees in the Garden: the tree of the knowledge of good and evil and the tree of life. The fruit of the latter was not explicitly forbidden, but awareness of the tree of life would come only after eating the fruit of the tree of knowledge. The fruit of the tree of life provided immortality, a foreign concept to those lacking the knowledge to appreciate it.

Adam was warned that, if he ate of the tree of knowledge, he would surely die. On the other hand, only the tree of life could provide immortality. Only by avoiding the tree of knowledge could he and Eve remain in their state of innocent bliss. Nevertheless, they were tempted by the serpent to eat the fruit. (Interestingly, serpents are symbols of circular or everlasting time in many primitive eastern religions, as in the case of the Jainist image of the serpent devouring its own tail.)[2] Both Adam and Eve tasted the fruit and became aware of their positions in the world. At this point God intervened to block them from tasting the fruit of the tree of life and from gaining everlasting existence. Adam and Eve were banished from the Garden and were condemned to lives of toil and the ever-present knowledge of mortality.

The fall was in some sense the true beginning of time and of history. With the knowledge of death comes the awareness of passing or linear time. We all experience this awakening when we first become aware of death. Hence the essential difference between childhood and adulthood: The child lives each day as if it were the only one; the adult plans for the future. As we grow older, we become more conscious of the passing of time. If we look upon the expulsion from the Garden of Eden symbolically, it represents the loss of innocence as the knowledge of death is revealed to us.

We may thus interpret the story of the fall as a chronicle of the passing from a circular to a linear, temporal perspective. By leaving the Garden of Eden, Adam and Eve abandoned the notion that every day is like every other. They were painfully made aware of the existence of death and of the ravage of time. Thus, history had begun for them.

The concept of time in the Bible is a linear, historical approach. Beginning with Genesis and the fall, the Bible is a chronicle of human deeds and divine intervention. All biblical events have dates of occurrence documenting a unidirectional world scheme. This stands in contrast to the Indian and Greek visions of the universe, which, as we have seen, present a cyclical approach to time, one without purpose or direction.

Divine events in the Hindu faith exist outside the realm of human time. They occur in a transcendental arena unwedded to the lives of individual people—that which Eliade has called "sacred time."[3] This corresponds to the Hindu picture of the world as recurrent. In such a vision, human history is inconsequential. Why should the behavior of the immortal gods have anything to do with the lives of humans in one particular cycle of a universe of infinite repetition?

The same holds true for Greek civilization. The ancient Greeks did not view religion as being in any sense historical. Mythological events were perceived as taking place outside the context of human activity; for, as Plato thought, the world as we perceive it is merely a shadow of the true, unchanging realm of the gods.

In contrast, the Jewish religion and the derivative Christian and Moslem religions view human history as part of a divine plan. Human events are perceived to occur but once and to quite often have a religious significance. That is why these religions are said to embody a linear picture of time unlike the cyclical time of most of the ancient world.

Judaism is obsessed with history and with the meaning of history. The laws of God are viewed as dynamic entities; they are revealed gradually rather than all at once. Someone born before the time of Moses would be expected by God to behave in one way, whereas those born afterward would be expected to uphold higher morals. In the Jewish world view, Mosaic law was provided by God at a precise moment in history. For Jews, human history has a purpose from its beginning at the time of the fall to its culmination in a messianic age of peace. The fact that there is a divine plan is never doubted, even though the details of the plan are seen as being revealed to humanity in a slow, fragmented way.

Christianity takes this point of view a step further. The Christian believes that the divine plan has already been revealed. The events of the Old Testament foreshadow the most important event in human history: the coming of Christ. It is clearly emphasized in the Christian faith that the birth, death, and resurrection of Christ took place only once at a precise moment in history: the age of Caesar Augustus. The possibility of multiple Saviors or multiple resurrections is a blasphemy against Christianity.[4]

According to Christian thought, the story of humanity is divided into four periods: the age of natural law, the age of Mosaic law, the age of grace, and the age of glory. The last age follows the second coming of Christ. The sequence of these periods makes sense only when presented in one particular manner; presented in another way, the meaning of history is lost. Hence, Christian time is linear and may be characterized by an arrow pointing from the past to the future.

In this manner, the revelations of the Old Testament prophets pointed to the coming of Christ. All that happened before his appearance served as preparation for this event. Those born before Christ had no possibility for salvation. Christ's coming was a critical and unique moment in history when, for the first time, human salvation became possible.

The message of human history is that, starting with the original sin of Adam and Eve, men and women have committed a series of sins and blasphemies against God. Left to itself, the human race will continue along the path of destruction until everything is destroyed in Armageddon, caused either directly or indirectly by human folly. Before the second coming of Christ and the end of time, the story of man is profoundly pessimistic because the road to destruction is irreversible. Only when human history is over will everything be set right as the Garden of

Eden is recreated. That is the picture of time painted by the early Christians, promoted by the apostle Paul, and refined by the fifth-century theologian St. Augustine.[5]

In his *Confessions,* St. Augustine divided the world into two spheres, the mundane world and the city of God. In his view, one should obey secular laws while living in the world, but at the same time devote oneself to the spiritual life. One must distance oneself from the evils of the world; for the earthly realm is doomed to destruction. Thus, Augustine's view of history is profoundly pessimistic. The only hope for salvation is through a complete faith in Christ, not simply through a belief that one can reform humanity. But in spite of his pessimism about human nature, Augustine finds hope in God's final judgment. Only after the Armageddon and the Last Judgment will the City of God be established on earth. Those who have been faithful will be reborn into this kingdom.

St. Augustine argued that Christianity and cyclical time are incompatible. Time began, according to him, at the instant of creation; it will end only when the kingdom of God is restored on earth. There can be no repetition of these events; eternal recurrence would rob them of their meaning.

From its earliest days, linear Christian time clashed with the cyclical time of the Greeks. The scholars of the Middle Ages had difficulty reconciling the idea of a divine plan that is revealed historically with the idea of an immutable God. For the Greeks, the divine presence is eternal and the universe is unchanging. For the Christians, the universe is developing and progressing toward a goal. Medieval scholars needed to resolve the apparent clash between the eternity of God and the providence of God. If the universe were created by a perfect God, it must be perfect. How could one justify any sort of historical development when it would only lead to imperfection?

In the short story "The Theologians," Jorge Luis Borges imagines what the medieval feud between advocates of circular time and Christian followers of linear time may have been like. Borges pictures a cult of followers of Plato, who use the wheel and serpent as their symbols. This cult advocates the concept of cosmic cycles. The worshippers of this dogma are defeated by the followers of the cross, who emphasize the linearity of time:

> Augustine has written that Jesus is the straight path that saves us from the circular labyrinth followed by the impious....Jesus was not sacrificed many times since the beginning of the world, but only once....Time does not remake what we lose; eternity saves it for heaven or hell.[6]

In spite of the rise of Christianity, the cyclical world view of the Greeks remained a persuasive force during the Middle Ages. Philosophy and science were strongly influenced by the work of Aristotle. It is quite understandable why cyclical time and a static model of the universe had such appeal during this time, given the lack of historical and scientific growth. History seemed to come to a standstill. Wars were fought and territories changed hands, but there appeared to be no rhyme or reason to these events. Empires rose and fell without any sense of historical progression. No new territories were discovered during the Middle Ages, so the world map remained fairly static. When space is static, time loses its significance in terms of measuring change. In other words, if space is unchanging, time does not have an arrow of direction.

Moreover, one can readily understand why circular time would appeal to the farming communities that characterized the Middle Ages. As we have seen, the cosmic cycle idea is strongly linked to agricultural society. The farmer is much more aware than the artisan of the seasonality of nature. At certain times of the year one must sow; at other times one can reap. People grow older, but nature seems to repeat itself. It was thought, by the superstitious medieval farmer, that perhaps, like the plants, people and events would return again and again.

It was only after the Renaissance, a period of artistic, scientific, and commercial development, that the idea of linear time began to predominate. This period represented a clear break with the static quality of the Middle Ages. Thus, people were ready to consider a dynamic concept of time. During this era of rapid change, the details of "God's plan" were again of much interest. In Christianity, the notion of linear time enjoyed a tremendous revival while the cyclic, ritualistic aspects of the faith were deemphasized.

Roughly corresponding to this period was the rise of Protestantism. The Reformation represented a shifting of Christianity away from its ritual aspects and toward its concept of historical redemption. We can see this clearly in the poetry of John Milton. Milton's epic poems *Paradise Lost* and *Paradise Regained*,[7] written during the seventeenth century, present a sequential portrait of the history of heaven, hell, and earth. Interestingly, these epics read like the chronicle of a great military conflict with a well-defined sequence of events: Satan and the evil angels do battle with the forces of heaven and then proceed to corrupt humanity, God's prized creation. Eventually the forces of God win out, as the story of man draws to a close. Thus, in Milton's work, human events are seen in the context of God's master plan — giving a meaning and form to time.

The hallmark of linear time is that the *order* of events is of great sig-

nificance. The meaning of Milton's chronicle is wholly dependent on
the sequence. Seeing the events in reverse order would not make sense.
In contrast, in the circular time scheme there is no perceived pattern of
change in the world. Therefore, the order of worldly events has no special meaning.

Compare, for instance, *Paradise Lost,* representative of the Christian
linear time approach, to Homer's *Odyssey,* representative of the Greek
circular time viewpoint. If one were to rearrange the events of Homer's
epic, it would be very difficult to tell that they were out of order. There
is a steady rhythm to the story that can be transposed without affecting
the plot: The hero, Odysseus, leaves home, sets out on an adventure,
then returns. He leaves home again, sets out on another journey, then
returns once more. His character doesn't change much throughout the
book. Although the Odysseyian chronicle is not circular in the strictest
sense, it is more or less atemporal in the way that its events lack an overall sequence and a definitive conclusion. On the other hand, Milton's
epic cannot be rearranged and still make sense. It has a clear conclusion, one that represents the end of human history.

There are many different approaches to linear time. If one believes
in linear time, one can assign a significance to the sequence of events
depending on one's belief. For example, a Miltonian Christian would
believe that the purpose of human history is the free acceptance of
Christ by man, leading to salvation and the restoration of paradise. A
Marxist would view human history as a dynamic process, rooted in economic factors, leading to pure communism. A pessimist might look at
the increase in the severity of human warfare and suspect that all events
point toward catastrophe. Yet, for all of these linear world views, the
meaning of history depends upon history's unidirectionality.

In the past few centuries, enormous changes have taken place in global politics, economics, and culture. These changes have been so dramatic that the idea of linear time has become increasingly influential.
Clearly, the two driving reasons for this revolution in human perception
are the tremendous growth of the human map of the cosmos in the
form of territorial expansion and spatial exploration and the related Industrial Revolution which radically altered the global economy.

Before the end of the fifteenth century, it was commonly believed by
European scholars that the world consisted of Europe, Africa, and Asia.
Europe was considered to be the center of the universe, and the other
continents had little significance. However, the last few centuries have
been a period of enormous conquest and territorial expansion. The discovery of America fully doubled the idea of the size of the world, and
exploration of Africa, Asia, and Australasia radically altered geograph-

ical perceptions. In a sense, the world grew by an enormous amount during this time.

The development of modern astronomy also had an enormous effect on the perception of the size of the universe. Once the idea of the enormity of space became known, the possibilities of expansion became limitless. The reach of humanity grew further and further. Finally, space came to be viewed as another human frontier. Astronauts assumed the role of modern-day Columbuses and Cooks.

Changes in the perception of space can have a strong influence on the time model favored by a society. When space is seen as unchanging, cyclical time is the favored model of the world; when space seems to be growing at an enormous rate and the frontiers of humanity are being pushed out further, the idea of progression is more appealing. Change, in the form of spatial expansion, generally happens in a unidirectional manner. The direction of change dictates the arrow of linear time.

The Age of Exploration of the sixteenth and seventeenth centuries was created by a search for new markets and raw materials; it coincided with the Industrial Revolution. During this period, the focus of society shifted from agriculture to industry and commerce. With the demise of agriculture as an important force in society, people lost their ties with the land, with the harvest, and with the cosmic rhythms. The idea of periodic regeneration, based as it was on the annual reappearance of plants, gradually lost importance. At the same time, the emerging capitalist society insisted upon growth and change as requisites for the accumulation of wealth and future development. Agricultural institutions can afford to be static, but commercial organizations need constant growth to generate new wealth. New resources must be found; new markets must be sought; and new laborers must be employed to keep the machines grinding away.

Along with industrial expansion came a growing fluidity in the social order. In a peasant society, it is generally impossible for the life of a farmer to change in any significant way. Caste differences are inviolable. Thus, it was natural for a peasant in a preindustrial society to perceive that every day was like every other day and that every era was like every other era. Bad times would occur, but they were bound to be followed by good times. For example, a drought might be followed by a time of great fertility and abundance. That led naturally to a cyclic, atemporal perception of events. In an industrial society, however, there is no rigid barrier to prevent upward mobility. Status is dictated by hard work and luck.

Therefore, after the Industrial Revolution the idea of change in one's life and in one's world came to be seen as natural and necessary. If people did not seek to improve themselves, they were lazy. If an organization or nation didn't develop, it was lagging behind the times.

The emerging Protestant religions stressed hard work and self improvement as signs of God's favor. The leading nations were seen as the bearers of God's plan. Naturally, Christianity became even more strongly wedded to a linear time scheme as the ideas of progress and change took hold. Just as Catholicism contains fewer traditional and repetitive aspects than Judaism, so Protestantism abandons many of the traditions of Catholicism. Therefore, of the three religions, Judaism is the least coupled to linear time. In Judaism, despite the notions that humanity is progressing and history is real, there is little sense of change or progress in the day-to-day life of the orthodox. Traditional Judaism emphasizes adherence to hundreds of laws designed to regulate daily life. The Jewish Bible is studied in a cyclic manner, with the same passages read on the same days of the year every year.

Unlike Judaism, the Catholic faith does not prescribe a complete daily ritual for the devout, but it does contain many repetitive aspects not found in the more radical versions of Protestantism. For instance, although both Christian dogmas insist on the historical nature of Christ's resurrection, the Catholic Church holds that Christ's blood and body return through the ceremony of transubstantiation. Most types of Protestantism reject this as a blasphemous notion and argue that Christ will not return again until the end of the world.

In Protestantism, and also in the liberal interpretations of Judaism and Catholicism, there is very little emphasis on tradition. Whatever traditions are practiced are considered to be nonessential aspects of the religion; the ceremonies are not considered sacred and immutable. The predominant notion of human progress and the importance of historical development have eliminated the earlier importance of those customs. The modern forms of all Western religions reflect the strong influence of the linear perception of time. As society has developed, the importance of ritual has continued to wane.

The Burden of History

Clearly, some price has been paid for the abandonment of the ancient dogmas. Undoubtedly, a certain amount of personal security has been lost in the transition. In his famous book *Cosmos and History*, Mircea Eliade laments the loss of the sense of connection to nature which stemmed from the replacement of religions based on cosmic cycles by the linear time of Christianity.[8] The peasant could find solace from natural disasters by knowing that good and bad times were all part of a natural cycle. Traditionally, if a series of misfortunes befell a certain society, it was looked upon as part of the downward trend of a time cycle. It was expected that the upward, or constructive, part of the cycle would soon follow. If time is linear, there is no such expectation. Things might continue to get worse and worse. Thus, history is a burden that all must carry.

The rituals in the peasant's life were linked to celestial archetypes; that is, they were seen as being handed down by the gods. The peasant would perform some function knowing that it had been repeated and would be repeated in the exact same manner for all eternity. That provided a sort of immortality for those who knew that their traditions were likely to be repeated in the same form forever. But once one accepts the idea of unidirectional change, one loses this solace.

Moreover, the belief in cosmic cycles has certain egalitarian aspects not embodied in the historicist's perspective that the world is changing. Historical change generally is governed by an elite group of men and women. Very few of us can achieve the power to affect history at all. On the other hand, each of us can practice certain traditions which give us another means of immortality; each of us can have a life that is sacred. The idea of historical progress essentially negates the power of traditions in which we can all share and replaces them with the concept that only a few of us can influence the world significantly.

Without eternal repetition, our beings are "light," our existence is insubstantial, as Milan Kundera asserts. Everything that we do is lost forever in the winds of change. Only a blessed few of us can have statues erected to ourselves, but even those monuments crumble. Finally, most ancient religions utilize cyclical time as an antidote to death. Through reincarnation, all creatures are repeatedly returning to earth in different corporeal forms. Religions based on linear time generally reject the possibility of reincarnation.

Eliade feels that the transition of human belief from circular time to linear time has produced a devastating, demoralizing effect. He likens it to the expulsion of Adam and Eve from the Garden of Eden. Adam and Eve lived in a paradise where every day was like every other—in other words, a cyclic cosmos. Their misery came from discovering death and the idea that change is possible. After that discovery, their lives were no longer holy and perfect. Similarly, by rejecting tradition and the notion of universal cycles, human life has lost the sacredness of eternal repetition.

The Christian solution to the problems resulting from the fall is a faith that, through Christ, someday paradise will be restored. This, in the Christian view, will occur at the end of human history. In other words, historical time will once again be banished and the timelessness of paradise will take its place. Eliade feels that this possibility is the only consolation for those who feel alienated by the linearity of time and the eventuality of death:

> Christianity incontestably proves to be the religion of "fallen man": and this to the extent to which modern man is irremediably identified with history and progress, and to which history and progress are a Fall, both implying a final abandonment of the Paradise of archetypes and repetition.[9]

By the beginning of the nineteenth century, agricultural society clearly was on the wane in Europe and in the newly independent United States. A spirit of progress and change was in the air as industrial development heralded an improved standard of living for most Europeans and Americans. As workers fled the farms and moved to the cities, it was wondered whether something may have been lost by abandoning the land and the agricultural tradition. Perhaps the simple life, with its comforting predictability, should never have been discarded? Perhaps there was a fall from a sort of paradise?

Nostalgia for agricultural society, and for the static time associated with it, appears occasionally in various benign and destructive forms. Generally this longing for a simpler time occurs when change is seen as having been too rapid or harmful. That sort of nostalgia can be found in Nietzsche's writings, with his revival of the concept of the eternal return. Since circular time can be associated with simplicity, perhaps Nietzsche's work was a reaction to the turmoil caused by the rapid changes of the nineteenth century. Another example of this fascination with eternal recurrence is the circular novel *Finnegan's Wake*, by James Joyce, in which the first sentence in the book is a continuation of the last sentence.[10]

The German historian Oswald Spengler, a leading advocate of a cyclical view of civilization whose works appeared in the 1920s, provides us with yet another example of this nostalgia. In his influential *Decline of the West*, Spengler likens a society to a biological organism. Just as each living thing matures, decays, and eventually dies, each civilization has a corresponding rise and fall. Every society experiences a period of cultural development followed by an age of stagnation and decline. Each new culture begins with the rise of a new mythology.

For instance, Indian culture began with the creation of the Vedic writings; Greek society, with the epic poetry of Homer and the Olympian legends. The "summer" of the culture, which is the next period, includes most of the characteristic philosophical writings. This is followed by the "autumn" years of greatest philosophical achievement.

Plato and Aristotle lived during the autumn of Greek culture. Finally, each culture passes into its "winter" years, a period in which the culture becomes a civilization. For Spengler, civilization represents an era of cultural stagnation — an age in which the distinctive features of the society are eroded and warped.[11]

Spengler stresses the similarities in the rise and fall of each culture. Since each society embodies a different set of values, it is foolish, according to Spengler, to proclaim that one set of values is superior to another. Therefore, one cannot claim that there are eternal truths. One must objectively study the general laws for the formation and propagation of these ideas without prejudice as to their merit. That is what is called philosophical relativism.

In equating the growth and decline of all civilizations, Spengler declares that there is no general trend in history. History cannot, in his view, be characterized by decay or progress: one cannot conclude that one's own era has any special features representing a general trend toward some goal.

After providing an overview of his approach to history, Spengler then documents how the Western world of his day was in a marked state of decline. The decline was characterized by the collapse of the German monarchy and its replacement by liberalism, socialism, and parliamentarianism. Unlike many of his contemporaries, Spengler associated democracy with decline and autocratic rule with high culture!

Advocates of circular time generally share a belief in a past golden age of glory, coupled with the feeling that their own age is somewhat lacking. They also share a strong hope that this era of glory will be restored. The rise of the Nazi movement in Germany is an example of the abuse of this nostalgia. One of the ways in which Hitler gained power was by exploiting the longing of the German people for a simpler, agricultural period. Although Hitler's actual policies were quite different, in his propaganda he emphasized his desire to restore medieval Teutonic values to German society. The idea of circular time played a major role in the Nazi myth, namely, the concept of the return of the Phoenix from its ashes. This was expressed in Hitler's vision of a thousand-year Reich. He argued that a new Germany would arise out of its devastating defeat just as it had in past cycles. It is no wonder that the detested emblem of the Nazis was a swastika — the ancient Indian symbol of circular time.

After the defeat of the Nazis, Hermann Hesse's masterpiece, *The Glass Bead Game*, appeared.[12] Hesse, who was strongly influenced by Buddhism, imagined a future which is much like the Middle Ages. In this imagined society, time is once again viewed as static when a new rigid order, devoted to philosophical speculation, controls the world. At the end of the novel, however, Hesse comes out against that sort of nostalgia and reveals a certain disdain for that rigidity.

A recent example of a yearning for paradise is the back-to-nature movement of the 1960s. Thousands of people sought a simpler life in agricultural communes. Once again there was a rejection of progress and change and a return to cyclical time myths. Indian philosophy enjoyed a revival, and the "dawning of the Age of Aquarius" was heralded as the beginning of an era of peace and love that would follow times of strife and bitterness. The interest in astrology during this time further indicated a popular belief in cosmic cycles.

It is interesting to ponder these questions: Why do nostalgic periods occur when they do? What unites periods of longing for simplicity? What drives people to hope that the past will come again in an eternal return?

One answer lies in the fact that these movements generally seem to occur after periods of devastating warfare (World War I, World War II, Vietnam). Over the last few centuries, warfare has grown in intensity and in cruelty. Generally, the greatest, and certainly the most destructive, changes in society have occurred through wars. National borders have been modified or erased; cities have been leveled; and peoples have been exterminated. It is during these gloomy periods that the terrors of history reveal themselves in their most direct form.

Naturally, there is a great hope that a period of construction will follow a war. Then the notion of circular time is most appealing. The terrible fear is that periods of destruction will continue, things will get worse and worse, and eventually Armageddon will occur, with the complete destruction of humanity as its outcome. Obviously, only those who believe paradise follows Armageddon would welcome the possibility of total devastation. Thus, the idea of temporal cycles provides a means of avoiding the horrible prospect of the end of time. It helps to alleviate the heavy burden of history.

Undoubtedly, the possibility of nuclear war has made the vision of the world's destruction seem all too real. We must now reckon with the possibility that human folly could bring time itself to an end. For if we associate time with change and development, clearly, after a nuclear devastation there would be no changes whatsoever. That would represent the ultimate proof of time's linearity, because a circle has no end. A nuclear holocaust would be the end of time for humanity.

There are those who might argue that time might continue in spite of the end of life on earth. Other planets might support intelligent life. In spite of time not continuing for the foolish earthlings, it would still flow for the more peaceful dwellers of the universe. Unfortunately, human folly is not the only way history could be brought to an end. The Industrial Revolution, besides triggering radical changes in culture and philosophy, revealed the seeds of the possibility of the end of the universe and the dusk of time through unavoidable natural circumstances. Science has discovered that, even if earth finds its way out from under the Damoclean sword of nuclear holocaust, the universe itself might someday be extinguished by its own internal mechanisms. In other words, it has become clear that the laws of physics governing the universe probably contain a built-in time limit.

The Law of Entropy

Until the nineteenth century, all of the known laws of physics were symmetric for both directions of time, for the past and for the future. The motion of objects could be explained by Newton's laws of mechanics. As we have noted, those laws do not distinguish between motion that is forward in time and that which is backward in time. Therefore, the Newtonian universe is a mechanism that behaves in a clockwork manner. There is no reason for it to run down: If the universe were grinding to a halt, that would seem to violate the reversibility in time of Newtonian classical mechanics.

During the nineteenth century, there was a great interest in engine design. The Industrial Revolution brought about a search for the perfect engine. It was felt that the efficiency of engines would continue to get better and better and engines would require less fuel to produce more work. All that was needed were better inventors.

One sort of nineteenth-century engine involved the use of steam power. A cylinder was filled with steam, which was then heated, causing the cylinder to expand. As the steam cooled, the cylinder would contract again. The motion of the cylinder was used to operate machinery.

In 1824, the French physicist Sadi Carnot, in his book *Reflections on the Motive Power of Fire*, addressed the problem of designing the perfect engine. Carnot considered an engine that develops motive power from heat flowing from a hot object to a cold object. Let us call the hot object, acting as the heat source, the *hot reservoir*. The cold object, acting as the place where the exhaust heat is sent, we shall call the *cold reservoir*. Heat from the hot reservoir would be utilized as work — to operate a piston, for instance. The unused heat would be expelled into the cold reservoir as exhaust.

For example, in a coal-powered plant, the burning coal would be used to generate steam, which would act as the hot reservoir. The steam would be used to operate a turbine, and the excess heat would flow into a cold reservoir (a river, for instance) as the steam cooled off. Carnot found that there was a strict limit to the amount of heat that could be converted into usable work. In other words, a certain amount of the heat energy would necessarily be wasted. It was impossible to eliminate this wasted portion of energy. The smaller the temperature difference between the hot and cold reservoirs, the more energy that would be wasted. That placed a strict upper limit on the efficiency of engines: some energy would always be lost in the form of exhaust. No matter how resourceful the inventor, he or she could never design an engine with 100 percent efficiency!

Thirty years later, the physicist Rudolf Clausius incorporated the work of Carnot into a new theory of heat and energy. To formulate his theory, Clausius defined two quantities: energy and entropy. Heat and work he considered to be two different forms of energy. In other words, heat could be converted to work and work to heat. Energy could never be gained or lost, but merely converted from one form to another. Clausius stated the so-called first law of thermodynamics in the following manner: "The energy of the world is constant."[13]

Entropy is a much more subtle concept than energy. For that reason, it has been perceived throughout the years as something intangible. Contrary to that perception, entropy can be precisely defined: it is a measure of the *incapacity* to obtain usable energy (work) from a certain process. For an engine, that depends upon the temperature difference between the hot and cold reservoirs. There might be a lot of *available* energy in some source, but it can best be converted into work only if the *entropy* is not too great.

For example, the sun's energy is so useful to us because its temperature is different from that of its surroundings. Therefore, it is a source of relatively low entropy. If the sun's surroundings were of the same temperature as the sun, the sun's energy would not be useful at all because the entropy would be too great.

The ocean contains vast reservoirs of heat, but most of that energy cannot be used. One would need a reservoir that was much colder than the ocean into which the heat would flow. A reservoir of the same temperature as the ocean would not suffice, for the entropy would not be low enough to permit transfer of the heat into work.

Clausius summed this up as the second law of thermodynamics, which he stated in the following manner: "The entropy of the world strives to a maximum."[13] In other words, there is a general tendency for entropy to increase in value. That means the amount of usable energy generally decreases while the amount of unusable energy continues to increase. Two bodies of different temperatures in contact with each other tend to exchange heat until they both have the same temperature. When the temperatures of the bodies are different, the entropy is low. Work can be done by utilizing this temperature difference (with a steam engine, for instance). As the temperatures of the objects approach each other, the entropy of the system increases. Finally, when the objects reach the same temperature, the entropy has reached a maximum and the temperature difference between the objects can no longer be exploited to produce work. The energy stored in the two bodies is no longer usable.

The process of temperature leveling and an associated increase in entropy occurs naturally. The opposite process, namely that of heat traveling from a cold to a hot object, does not occur naturally. Thus, events occur in the direction of increasing entropy. Not every process results in entropy increase, but any local decrease in entropy must be accompanied by a corresponding increase in entropy elsewhere.

In the scheme of the universe, that means that eventually all temperatures would become equal. At that point, the entropy of the universe would reach a maximum. From that time on, no work could be done in the whole universe. No engines could be created to use any of the energy left in the cosmos. Therefore, no change would be possible anywhere in the universe. This final state is called the heat death of the universe.

The increase in entropy provides a natural arrow of time. Processes that occur forward in time are associated with a general increase in entropy. A closed process in which entropy decreases could not run in the forward direction of time's arrow. Therefore, one can associate the striving of entropy toward a maximum with the flow of time.

Let's see how increasing entropy implies that time has an arrow. Consider the mixing of two vats of water, one of which is initially very cold and the other very hot. This great temperature difference can be associated with a low value of entropy. If the water from the two containers is then mixed, eventually the temperatures in the two vats reach the same value. As that occurs, the entropy increases to a maximum value.

We can associate an increase in entropy with the forward arrow of time. To see how, imagine running the whole process backward in time. Initially, we would see a vat of lukewarm water. Suddenly, the water would separate into two vats. The water in one vat would be increasingly hot as more heat flowed into the container; the water in the other vat would be increasingly cold. As the temperature difference increased, the entropy would have decreased.

. It is obvious that we would never see this process occurring naturally. Water never divides into hotter and colder parts. If we could see something like it occurring, we could only assume that time was flowing backward. In nature, we observe only the entropy-increasing, forward-time process. That is why entropy is referred to as an arrow of time.

Another example of entropy increase is the shattering of glass. Shattered glass has a higher entropy than unshattered glass; so when glass is shattered, entropy increases. However, if we saw a sheet of glass assembling itself from a shattered state to an unshattered state, we would necessarily conclude that the film was running in reverse. Entropy would decrease in that case and time would be said to flow backward.

One might wonder what broken glass has to do with heat engines. Breaking glass does not represent a transfer of heat in the same way that mixing vats of hot and cold water does. Nevertheless, it represents an increase in entropy in a broader sense of the word.

Entropy can be defined in the most general sense as a measure of *disorder*. Thus, the second law of thermodynamics can be stated as the proposition that processes occur in the direction of increasing disorder. A natural process by which a closed system spontaneously becomes more ordered cannot occur, so the disorder in the universe strives toward a maximum. In the example of the broken glass, an ordered state (unshattered) becomes a disordered state (shattered). The reverse situation is impossible; it would represent motion in a direction opposite to that of time's arrow.

The second law of thermodynamics radically altered the status of time in physics. Newtonian time is static; there is no way to distinguish processes which occur earlier or later or forward or backward in time. The Newtonian universe could never run down; all the planets would continue in their orbits forever. Therefore, Newtonian time is closely related to the cyclical universe models of Plato and Aristotle. On the other hand, the entropy law introduces a preferred direction of the flow of time. The universe can no longer be considered static; instead it is constantly evolving toward states of greater disorder and higher entropy. It is thus possible to distinguish between earlier and later times. The later times correspond to an increased level of entropy. The flow of time forward is associated with an increase in disorder. Clearly, this is a linear model of time.

Moreover, the law of entropy enables us to imagine an end to time: the heat death. Picturing the universe after the heat death, one cannot speak of the flow of time, since no natural processes could occur. Time itself would come to a halt. Needless to say, many nineteenth-century philosophers and scientists found this prospect to be profoundly depressing.

Not all theorists have accepted the necessity of the universe ending in a heat death. Some, notably Milne, have argued that the second law of thermodynamics does not apply to the universe as a whole. Others, such as the German physicist Boltzmann, have asserted that entropy increase may be a phenomenon associated only with our section of the universe.

According to Boltzmann, one cannot talk about entropy or the second law of thermodynamics on a microscopic level. All physics on the atomic level is completely reversible, since particle motion is completely determined by Newton's laws of motion. However, on a macroscopic (human) level, one can define entropy as a statistical measure of disorder. To do so, one considers all possible arrangements of molecules, given a certain temperature or amount of energy. Some of these arrangements are much more likely than others. Entropy can thus be defined as a measure of how likely a certain arrangement is.

For example, if one considers all possible arrangements of grains of sand, it is very unlikely that a random selection of one of the arrangements will be a sand castle. That unlikely arrangement would correspond to a low entropy state. A much more likely configuration would be an amorphous pile of sand. The pile of sand would be said to be in a state of high entropy. In other words, there are many more ways in which grains of sand can be arranged in a shapeless manner than in the form of a sand castle. Entropy is thus a measure of the *lack of uniqueness* of a configuration.

Boltzmann argued that in some parts of the universe entropy would increase and in other parts decrease. In the sections in which entropy decreased, time would flow in the opposite direction. The average entropy of the universe would remain approximately constant. Any increase or decrease in entropy in some part of the universe could be explained by local fluctuations in the value. Therefore, one could not define an arrow of time for the whole universe, and there wouldn't necessarily be a heat death. Boltzmann, however, did not strictly rule out the possibility; he merely considered a heat death to be unnecessary.

In order to explain why the entropy in our section of the universe is increasing, Boltzmann used an approach called the anthropic principle, which states that the physical conditions of the early universe, or at least our part of the early universe, must have been such that intelligent life was possible. The argument goes as follows: If a part of the universe did not have those conditions, no one would be alive to observe the physical phenomena. The reason we are in a section of the universe that has low entropy is that only in low-entropy regions can the physical and biological processes necessary to produce life take place. In other sections of the universe, life would not be able to form, so there would be no intelligent observers. Thus, the argument goes, we are here in a low-entropy part of the cosmos because if we were in another section of the universe, we wouldn't exist as intelligent creatures and therefore wouldn't realize our predicament! There has been no experimental evidence to support or to completely disprove Boltzmann's hypothesis, so it has continued to be a matter of some controversy.

Many attempts have been made to look for "loopholes" in the second law of thermodynamics. Perhaps the most famous of these is "Maxwell's demon" hypothesis. Soon after the publication of the articles by Clausius, the physicist James Clerk Maxwell proposed a way in which entropy could be reversed. He imagined the existence of a demon of atomic size. The demon would sit at a gate between two containers of liquid. At the start, the contents of the containers would be at equal temperatures, i.e., a state of high entropy (complete disorder). The demon would screen all of the liquid's molecules passing through the gate and it would open or close the gate in a selective manner. All of the molecules of high speed would be sent to one vat, and all of the molecules of lower speed would be sent to another vat. Gradually, the temperature of the vat with the high-speed molecules would increase, because temperature is related to average molecular speed. The temperature of the other vat would decrease. In that manner, order would arise from disorder and entropy would decrease which would be a contradiction of the law of entropy.

The flaw in Maxwell's argument is that the demon must make observations in order to determine the velocities of the molecules of the liquid. To do so, the demon would require a sort of miniature lamp. The light from the lamp would have to be precisely focused in order to make detailed observations, so the light would start out in an ordered (low-entropy) state. After the observations, the light would be scattered and hence disordered. The increase in the disorder of the light would more than offset the decrease in the disorder of the molecules. Therefore, the total entropy of the system would increase, rather than decrease! The second law of thermodynamics would not be violated.

Another argument against the second law involves the Poincaré recurrence hypothesis. According to this conjecture, one can use Newton's laws to calculate the future positions and speeds of all of the particles in the universe. If the number of those particles is finite, then all of the molecules should eventually return to the same state. (This assumes that the size of the universe is constant, an assumption that the Big Bang theory has rendered obsolete.) If all of the particles were to return to their original positions, then any ordered state would eventually become an ordered state again, even after becoming disordered along the way. That would contradict the second law and provide for a cyclical, rather than linear, theory of time. The reader might recognize this argument as similar to that which motivated Nietzsche to propose his theory of the eternal return.

The major drawbacks in the above line of reasoning are as follows: First of all, the assumption that the size of the universe is fixed. If the universe is expanding, then the particles needn't ever return to their original positions. Second, even if the particles do eventually return to their original positions, there is no reason for them to return to their original speeds. Last, even if one allows for such a recurrence, the amount of time for it to take place is astronomically large, much larger than the time scales that one considers in cosmology.

Most physicists now believe in the universal validity of the second law of thermodynamics. Moreover, it is generally thought that the direction of entropy increase, and hence the direction of the arrow of time, is the same for the entire universe. Thus, in our current picture of the cosmos, one can conclude that the universe started out in a state of low entropy and that it will eventually reach a state of maximum entropy or heat death. This provides a clear distinction between the concepts of "before" and "after," one that is not provided by Newtonian mechanics.

Arrows of Time in Physics

The discovery of the entropy law stimulated a search for other arrows of time in physics. It was considered remarkable that physics on a small scale is totally time-reversible, whereas physics on a large scale is asymmetric with respect to time. Therefore, physicists have looked for time reversal asymmetry on a microscopic scale as well. The search has yielded several examples in particle physics in which there may be a difference between the backward and forward time pictures.

One example of time asymmetry lies in the theory of radiation. During the nineteenth century, a comprehensive study of electrical and magnetic phenomena yielded a rather remarkable and complete theory of electromagnetism. By the theory, electricity is carried by charged particles called electrons. When electrons oscillate, they produce electromagnetic waves. These waves can be associated with visible and invisible light radiation.

According to the laws of electromagnetism, there is no difference between an electron absorbing or radiating light. Radiation might be thought of as a time-reversed version of absorption. Electromagnetic theory, like Newtonian mechanics, does not distinguish between forward and backward time directions. Theoretically, radiation and absorption should occur with equal probability. In practice, nothing of the sort occurs. If an electron starts to oscillate, radiation of light is the only possible result. The time-reversed scenario is impossible. This is an apparent time asymmetry in particle physics.

Another situation in subatomic physics that has been mentioned as an argument for unidirectional time is the case of time symmetry violation in some types of kaon decay. (A kaon is an unstable subatomic particle that decays into other subatomic particles.) This Nobel prize-winning discovery by the American physicists J. W. Cronin and Val Fitch developed from a series of experiments during the early 1960s. It was found indirectly that the time inverse of the kaon decay process takes place at a rate different from that of the forward-time process. That provides a means of distinguishing before from after on a microscopic, as well as macroscopic scale. But it is not known to what extent time reversal asymmetry can be found in the subatomic world.

It is unclear what relationship these microscopic arrows of time have with the large-scale arrow connected with entropy increase. This problem has been the focus of much controversy; no one has developed a cogent explanation of it. Moreover, it is of considerable interest to determine the nature of the connection between these arrows of time and the intuitive notion of time flow. The latter phenomenon is usually referred to as the psychological arrow of time; it represents the feeling that one is traveling from past to future. No one has provided a satisfactory explanation of whatever relationships may exist between these arrows.

Finally, there is another arrow of time with which we must reckon. It is the so-called cosmological arrow of time, pointing in the direction of the expansion of the universe itself. A truly comprehensive model of the nature of time would determine the common origin of all these arrows.[14]

The Big Bang and the Big Crunch

In the first chapter we discussed the astounding discovery that the distant galaxies in the universe are moving away from ours and we mentioned the so-called Big Bang and steady-state theories as possible explanations of the phenomenon. Because of the progress made in cosmology over the past few decades, it has been concluded that the universe has evolved from a point object and has been, and will probably keep on, expanding outward. The expansion of the universe provides another natural arrow of time: the cosmological arrow.

The debate between advocates of the Big Bang and those of the steady-state theories of the universe formed the center of cosmological inquiry during the 1950s and early 1960s. Big Bang theorists argued that the universe emerged as a fiery ball of matter sometime in the past; the steady-state proponents vehemently asserted that the universe did not have a beginning, and they utilized other methods to explain the recession of the galaxies.

In many ways, this dispute paralleled the religious and philosophical debates between advocates of cyclical and linear time. Some thinkers have found the idea that the universe could have a beginning or an end to be distasteful. Others have found the idea that the universe may last eternally, and that events may repeat themselves, to be horrifying.

Curiously, this seems to continue the medieval dispute over the nature of God. The Greek approach, which sets the activities of the deities in a place outside the natural world and in a time outside human experience, seems best suited for a steady-state picture of the cosmos in which there is no instant of creation. On the other hand, the Christian linear time approach seems to parallel the notion of the creation of the universe as an event that is fixed in time. It allows for the possibility that God has somehow decided the initial conditions of the universe while setting it in motion through the Big Bang. Admittedly, this is a rather broad and unorthodox interpretation of Christianity. However, the Big Bang theory, unlike its alternative, does provide a mechanism by which one can believe in God as creator while still accepting modern science.

Fortunately, the dispute between these conflicting cosmologies has been settled. Clear experimental evidence supporting the Big Bang conjecture has been found. Today almost all cosmologists accept the premise of a fiery origin of the universe.[15]

The experimental verfication of the Big Bang theory took place over the last few decades. In 1964, two scientists at Bell Laboratories, Arno Penzias and Robert Wilson, noticed something unusual while examining a satellite dish antenna. A uniform background of signal noise was being picked up by the device, which Penzias and Wilson discovered had the same intensity in all directions. After eliminating all possible earthly and galactic origins, researchers concluded that the signal emanated from the residual radiation of the Big Bang itself. That would explain the uniformity of the background signal, since one can trace the origin of all points in space back to the initial fireball. Moreover, the background radiation was found to have almost exactly the wavelength that the Big Bang theory predicted, which would make it highly unlikely that it came from a different source. This Nobel prize-winning discovery served to tip the balance of the cosmological debate strongly in favor of the Big Bang model.

Other astrophysical evidence tends to support that position. The amount of helium contained in the universe suggests that the element must have been created in the initial fireball; the steady-state model cannot account for the helium abundance. Also, recent observations of the most distant parts of the universe have detected primordial objects of galactic dimension, and there is evidence that they are in an earlier stage of evolutionary development. Since the light from them takes so long to travel to us, we can assume that these objects are the oldest structures in the observable universe. That supports an evolutionary model of the cosmos and a rejection of the static approach of the steady-state model.

There still is much controversy over the nature of the end state of the universe, mainly because all the evidence is not yet in. We have mentioned an alternative: the oscillatory model of the cosmos, in which the universe recontracts to a point after its initial expansion phase. In this scheme, time is cyclical—there is no beginning or end to the universe. The other possibility is that the universe will continue to expand forever.

General relativity provides us with a means of determining whether or not the universe will oscillate. However, in order to determine which possibility is the most likely one (according to Einstein's theory), one needs to know the mass of all of the matter in the universe. According to current estimates of the amount of matter in the universe, it is more likely than not that the universe will continue to expand forever. There does not seem to be enough mass to cause a recollapse. Critical measurements of this quantity were made during the 1970s, notably by the CIT and University of Texas researchers J. Richard Gott, James Gunn, David Schramm, and Beatrice Tinsley. During the 1980s, scientists revised the estimate on the basis of proposals of new elementary particles and so-called dark (invisible) matter. However, the figures still seem to be too low to support a recollapse because most of the particles postulated to account for the missing mass have not been found experimentally.

If one believes in an oscillating universe model but still associates the flow of time with the expansion of the universe, one arrives at a strange conclusion, namely, that time would flow backward during the contracting phase. We have mentioned that Stephen Hawking has been an advocate of that position. More recently, however, in developing a comprehensive theory of the connections between time's arrows, Hawking has revised his opinion of this matter.

In *A Brief History of Time*, Hawking presents a theory of the relationship between what he calls the three fundamental arrows of time: thermodynamic (entropy increase), psychological, and cosmological.[16] According to him, we perceive time to flow in one direction because it is the direction of information storage. Our memories accumulate knowledge in just one way: from the past toward the future. The feeling that time is passing comes from an increase in the amount of information stored in our brains. That defines the psychological arrow of time.

Why is it then that the direction of the psychological arrow of time is the same as that of the thermodynamic arrow, Hawking inquires? In order for our bodies to function, and therefore for our brains to function, we need to eat. When we eat and metabolize our food, we are converting energy that is relatively ordered to energy that is more disordered. That is an entropy-increasing process that requires the presence of a reservoir of ordered energy. Thus, Hawking shows that the direction of entropy increase must be the same as the direction of life processes. One of those processes is memory accumulation. Hawking thereby concludes that the direction of entropy increase, which is depicted by the thermodynamic arrow, must be the same as the direction of memory increase, which is indicated by the psychological arrow.

Now that he has provided an argument for equating the thermodynamic and psychological arrows of time, Hawking proceeds to argue that the direction of entropy increase must be the same as the direction of time indicated by the expansion of the universe. First of all, life can develop only when there is a fairly low amount of entropy in the universe. When there is too much entropy, there is not enough ordered energy for life to flourish. Thus, one might expect life to flourish only in a fairly early phase of the universe, before the approach toward heat death has brought most processes to a halt. Assuming that the universe expands and then contracts, this early phase occurred during the era of expansion. Therefore, the direction of life processes, which must be toward an increase in entropy, coincides with the cosmological arrow of the expanding universe.

Hawking concludes that the answer to the question, "Why is the universe in its expanding phase?" is "Because we exist." If it weren't for the formation of life, which requires a low-entropy stage of the universe (identified with the expanding phase), human observers would not be around to tell the story. This is a fine example of the use of the anthropic principle to explain a law of physics by the fact of our existence.

What would happen during the recontraction of the universe, if it were to occur at all? Clearly, the arrow of time associated with the expansion of the universe will reverse itself. On the other hand, the arrow of time associated with entropy increase will continue to point in the same direction, if it existed at all. (After the heat death, there will be no thermodynamic arrow of time, for entropy will have attained its maximum value. Life will have died out long before the universe recontracts to a point.) Thus, the last stage of the universe will be a time in which there is little or no change as the usable energy in the universe dwindles away to nothing.

Roger Penrose, a British mathematician and cosmologist, has proposed an alternative explanation for the cosmological law of entropy increase.[17] Penrose associates the entropy of the universe with a mathematical object (called the Weyl tensor) utilized in Einstein's theory of general relativity. This mathematical construct is required to be zero at the beginning of the universe, but it has a larger value as the universe develops. In this way, Penrose argues that entropy increase is "built in" to Einstein's theory of gravitation. Penrose feels that the anthropic principle argument cannot adequately explain the second law of thermodynamics and the flow of time. Instead, he hopes that the concept of entropy increase can be incorporated into a unified picture of the laws of gravitation.

Hawking and Penrose, along with the bulk of all modern physicists, feel that the second law of thermodynamics is immutable. Entropy will continue to increase until it reaches a maximum. Modern-day cosmologists assert that we have no hope of escaping the eventual heat death of the universe, just as Christian theologians predicted that the world will end in Armageddon. Today, both fundamentalists and "orthodox" physicists agree that human history will most likely be brought to an end some day. They would, however, disagree on the nature of the timelessness that will follow the end of history.

An interesting fusion of thermodynamics and religion is described in the short story, "The Last Question," by Isaac Asimov. This story details the heat death of the universe and the futility of human attempts to escape it. Finally, the universe meets its demise. At this point, God creates the world all over again in an enormous explosion, while proclaiming "Let there be light!"

What Asimov seems to suggest here is the possibility of a reconciliation between the Biblical stories of creation and Armageddon and the modern physical theories of the Big Bang and the heat death. Although Asimov's intention here is simply to provide a good twist to a science fiction story, his suggestion of including the action of God as a component of physical theory has its parallels in the writings of many modern cosmologists.

Modern physics, by proclaiming a beginning and an end to time, seems to create some room for the possibility of divine intervention. Generally, this involvement is limited to choosing the initial physical parameters of the universe. In other words, God's role is to specify the set of constants that are required for certain physical laws. It is curious that physics now attempts to define the boundaries of God's powers: God has the task of closing gaps in cosmological theory!

It is interesting to examine modern physical concepts regarding time's end. In the current cosmological jargon, the end of the universe is referred to as the Big Crunch. There are two possible scenarios for this collapse. The distinction between the possibilities is the question of whether or not the universe *as a whole* will contract to a point. In both scenarios, however, sections of the universe will collapse to points; in fact this may already have occurred.

The mechanism for the demise of the universe is closely related to the nature of star death. Stars can die in a number of ways, depending upon their size and intensity. Stellar extinction comes when the nuclear reaction inside a star can no longer take place; in other words, when the star's entropy has reached a maximum. The star can no longer utilize any of its stored energy. When that happens, it becomes energetically favorable for the star to either shrink or explode (nova).

What happens next depends upon the size of the star. The small stars, called white dwarfs, continue to shrink while giving off light energy. White dwarfs glow for a time and then completely burn out. What is left is a fossil star, called a black dwarf, which travels through the universe as a cold dark relic.

Medium-size suns end up as neutron stars. The distinguishing feature of these objects is that all of the atoms in them have collapsed under enormous gravitational pressure. What is left is an extremely dense substance, called neutronium, consisting of closely packed neutrons (electrically neutral subatomic particles).

The biggest stars have the most dramatic endings. (This is true in Hollywood, as well as in the heavens!) These stars continue to collapse until they are incredibly dense. Because they are so massive, their presence distorts the structure of space-time around them. This is an application of Einstein's theory of general relativity—that matter alters the geometry of space-time. These objects are called black holes; they are named for the fact that not even light can escape from them.[18]

The inside of a black hole is very strange. According to theory, there is a region inside the black hole where space and time reverse roles. In these regions, presumably, one can move freely in time but not in space. (This is highly speculative.) At the very center of a black hole is a mathematical anomaly called a space-time singularity. These objects are predicted by the mathematics of the theory, but no one knows what to make of them physically.

According to theory, if an object reaches a singularity, it just disappears. Space and time simply vanish, which is what pre-Columbian mapmakers predicted for sailors reaching the edge of the earth. Needless to say, the existence of the singularities is disquieting for most cosmologists. Many theoretical researchers, including Hawking, are working on eliminating these objects from gravitational theory.

Eventually, all of the stars in the universe will become either stellar remnants or black holes. New stars will form, but the second law of thermodynamics mandates that the amount of usable fuel for the formation will dwindle. Finally, there will be no more usable energy left. As the stars die, any lifeforms on the planets orbiting them will no longer be able to survive, for the amount of planetary fuel also is limited.

Stars in the vicinity of black holes may be absorbed by them. Eventually, entire galaxies may become black hole structures. Black holes have a tendency to combine, and when two of them merge, the resulting object has a larger surface area than the original two.

A second law of thermodynamics has been formulated for black holes. According to Hawking and Bekenstein, the area of the event horizon of a black hole can be associated with the hole's entropy. The event horizon is the boundary between the normal and anomalous regions of space-time. Because of the law that entropy strives to reach its maximum, the surface area of the boundary has a tendency to increase.

Until recently, it was thought that neither matter nor energy could escape from a black hole, but the latest theories of gravity postulate that energy slowly leaks out of a black hole through a process known as quantum tunneling.[19,20] Eventually, all of the matter and energy inside a black hole is dissipated. What is left is called a naked singularity — a rip in the fabric of space-time.

The final state of the universe, according to current theory, will consist of nothing but cold stellar remnants, dissipated energy, and naked singularities. If the universe eventually recontracts, all these objects will merge and form one final singularity. Otherwise, if the present cosmological expansion continues, these stellar remnants will simply drift apart or disappear into the singularities.

Clearly, there is no escape clause in this scenario that would allow for the possibility of the survival of humanity. All that has ever been created would perish in such a heat death and Big Crunch collapse of the universe.

The Significance of Entropy Increase

All the events that we have just described in our account of the end of the universe would not take place for billions of years. If that is the case, the reader may wonder what their present-day significance is.

One can draw an analogy between the life of a human being and the life of the human species. At some point in a person's life, there is a realization that human existence must end in death. One of the ways in which people come to terms with their own mortality is the comfort that the species will survive. Until this century, science did not explicitly rule out that possibility.

Presently, science cannot provide any method of assuring the survival

of the human race. Therefore, the fact that the human race is doomed
to extinction provides even more cause for anxiety. That is especially
true if one does not believe in a deity or a supernatural realm. Jacques
Charon describes the depressing aspects of that prospect in his work
Modern Man and Mortality:

> What is particularly distressing for those who seek a "death-proof"
> meaning and purpose of life without taking refuge in the "other"
> world is that science actually confirms the conclusion of meaningless-
> ness to which the fact of death led man long before science had
> shaken his belief in immortality and disclosed the insignificant place
> of his abode in the universe.[21]

Charon asserts that the discoveries of the law of entropy and the Big
Bang have had a demoralizing influence. Not only do these discoveries
provide an arrow of time for the universe but the arrow that they suggest is one pointing toward a meaningless, empty, depressing future. It
is no wonder that Nietzsche and many other thinkers have embraced
the doctrine of eternal recurrence and that Fred Hoyle and other cosmologists have argued for a static, steady-state model of the universe! It
is difficult to accept the notion of the end of all things.

It is interesting that the law of entropy appeared during a period of
tremendous industrial growth and that the idea of black holes and a Big
Crunch came about during the dawning of space exploration. In both
cases, science has suggested limits on the power of humanity during
times when it appeared that there were no limits. One can no longer
assume that the invention of more efficient machinery will solve the
problem of inadequate natural resources, nor can one successfully argue for territorial expansion as a cure-all. Eventually, there is nowhere
to go; even the colonization of space is limited by the prospect of stellar
and galactic extinction. One must face the finiteness of one's own resources.

In *Entropy — A New World View*, the economist Jeremy Rifkin asserts
that humankind has not adequately dealt with the limitations implied by
entropy increase. Rifkin argues that the history of industrialization is
one of thoughtless, irreversible exploitation of natural resources. As one
resource is used up, a crisis develops until a replacement can be found.
In most cases, the replacement is less energy-efficient than the original
resource. In that manner, our society is rapidly robbing the earth of all
its usable fuel sources. Instead of acting in a manner designed to slow
down the global entropy increase, we are in fact hastening it.

For example, in switching from walking to horse-drawn transport to
trains and then to the automobile, humankind has opted for transpor-
tation that is more and more inefficient and hence entropy-increasing.

Farming methods of today require much more energy per crop yield than did those of the Middle Ages. Energy sources have become dirtier, more costly, and increasingly inefficient. More and more energy is needed to extract the metals which are critical for our industries.

Rifkin advocates a global recognition of the limitations imposed by the second law of thermodynamics. He suggests that, although there is no real long-term solution to the problems imposed by entropy increase, we can attempt to push back the time limitation by carefully conserving our resources and by reducing our dependence on inefficient technologies. For example, solar energy provides a source of power that is renewable and nonpolluting. An increased reliance on solar power would postpone the exhaustion of our resources. Rifkin notes that:

> In the past scholars have equated entropy with the final Heat Death of the solar system and then concluded that it is not of great concern to human life, since that eventuality is so far off into the distant future. In contrast [one should] focus attention on entropy as a process rather than a final state.[22]

Present-day humanity has unprecedented power to control its own duration in time. Through nuclear warfare or massive environmental exploitation, history could be brought to a close tomorrow. On the other hand, through conservation and planning, the end of civilization might be postponed for generations, possibly even approaching the time limits imposed by nature.

Perhaps the clearest symbol of human power over time is the clockface emblem of the Union of Concerned Scientists. Its minute hand is only a few minutes away from midnight and it is adjusted whenever global events seemingly bring us closer to or further away from nuclear Armageddon (midnight). It provides a bold reminder of the power that humankind has over its own destiny.

The Fall from Grace: The Downward Slope of Time

Since the Industrial Revolution, humanity has become increasingly alienated from the soil and from the seasons. The "simple life," with its rituals and rhythms connected to agricultural cycles, has been largely abandoned in most of the western world. The traditional notions of renewal and harmony have been replaced by a drive toward expansion and development. In all this, a certain type of innocence has been lost. For the first time, people must deal with a rapidly changing culture in which values and traditions are shed whenever they do not fit emerging economic and social frameworks. Without a sense of continuity, and even lacking the comfort of the certainty of species survival, humankind has walked away from a stable agrarian safehold toward an unknown future of either progress or destruction. The fruit of the tree of knowledge has been eaten; now there is no turning back.

One can hardly say that preindustrial society represented any sort of paradise. Surely, peasant life was short and harsh. Yet, as Eliade points out, early cultures provided a means to immortality that somehow has been lost. Through ritual repetition linked to celestial archetypes, people could feel a sense of connection with the infinite. Circular time provided a reassurance of continuity; it was believed that after every natural disaster there would come a period of growth and fortune.

The meaning of time has changed in both religious and scientific cosmological models. The universe is no longer seen as static and clockwork; the "city of man" is no longer considered to be permanent. Time is now viewed as dynamic rather than static, as historical rather than ignorable.

From Judaism to Christianity and Islam, religious philosophies have developed an increasing sense of history and purpose. Following in the tradition of St. Augustine and Milton, the earthly realm is seen as being corrupt and distorted. In this view, however, human history serves as a preparation for the new paradise that will follow the complete destruction of the secular kingdom of man. Traditional Christian linear time has an arrow which points in the direction of the decay of civilization.

Science also has come to the conclusion that time has an arrow. In fact there are several, possibly related, arrows of physical time. The arrow of increasing entropy serves as a counterpart to growing industrialization, as a reminder of the limits of our resources. No longer can we expect technology to provide a panacea for problems arising from the depletion of fuel sources. Ultimately, the second law of thermodynamics provides a time limit for the universe. There will come a time when, because of the lack of available usable energy, no more change is possible. When the heat death occurs, entropy will have reached a universal maximum.

Cosmological theory provides us with another arrow of time: the growth of the universe itself. According to the current model, the universe has expanded outward from a pointlike origin. The universe might end in recontraction, or it may in fact continue to expand forever. In either case, eventually most of the universe will consist of black holes and burnt-out stars.

In this chapter, we have explored many aspects of what one might call the downward slope of time. Essentially, this is the point of view that the universe is not static or renewable but is in fact running down. As we have seen, there is considerable evidence for this hypothesis, in both the world of science and the world of human affairs.

In the next chapter, we shall consider a different, more hopeful, point of view: the theory that the world is evolving toward order, not chaos. That is what can be called the upward slope model of time.

References

1. Jorge Luis Borges, "The Immortal," in *Labyrinths*, New Directions Publishers, New York, 1962.

2. Ruth Reyna, "Metaphysics of Time in Indian Philosophy," in Jiri Zeman (ed.), *Time in Science and Philosophy*, Elsevier Publishers, New York, 1971.

3. Mircea Eliade, *Cosmos and History—The Myth of the Eternal Return*, Harper and Row, New York, 1959.

4. Louis Gardet, *Cultures and Time*, The Unesco Press, Paris, 1976.

5. Geza Szamosi, *The Twin Dimensions*, McGraw-Hill, New York, 1986.

6. Jorge Luis Borges, "The Theologians," in *Labyrinths*, New Directions Publishers, New York, 1962.

7. John Milton, *Paradise Lost*, The Odyssey Press, New York, 1962.

8. Mircea Eliade, *Cosmos and History*, op. cit.

9. Ibid.

10. James Joyce, *Finnegan's Wake*, The Viking Press, New York, 1947.

11. Oswald Spengler, *Decline of the West*, G. Allen and Unwin, London, 1928.

12. Hermann Hesse, *Magister Ludi—The Glass Bead Game*, Bantam Books, New York, 1969.

13. S. Carnot and R. Clausius, in E. Mendoze (ed.), *Reflections on the Motive Power of Fire and Other Papers on the Second Law of Thermodynamics*, Dover, New York, 1960.

14. Richard Morris, *Time's Arrows*, Simon and Schuster, New York, 1984.

15. Edward R. Harrison, *Cosmology: The Science of the Universe*, Cambridge University Press, New York, 1981.

16. S. W. Hawking, *A Brief History of Time*, Bantam Books, New York, 1988.

17. Roger Penrose, "Time—Asymmetry and Quantum Gravity," in Isham, Penrose, and Sciama (eds)., *Quantum Gravity 2*, Oxford University Press, New York, 1981.

18. John G. Taylor, *Black Holes*, Avon, New York, 1973.

19. S. W. Hawking, *Commun. Math. Phys.*, **43**:199.

20. J. D. Bekenstein, *Phys. Rev.*, **D7**:2333.

21. Jacques Charon, *Modern Man and Mortality,* Macmillan Publishing Co., New York, 1964.

22. Jeremy Rifkin, *Entropy—A New World View,* Bantam Books, New York, 1980.

3
The Road Slopes Upward

The Idea of Progress

Of all the models of time's direction, the idea of progress is the most recent. The circular view of time can be traced back to the beginning of agricultural civilization and the idea of degeneration can be said to have originated at least as far back as the early Christian philosophical writings, but the progressive time viewpoint is only several centuries old.

The idea of progress provides a distinct arrow of linear time. Advocates of the progressive ideal argue that the world is evolving toward a state of greater order and complexity. In contrast to the idea of eternal repetition, the upward slope model of time represents a unique progression of events in a clearly identifiable sequence. Unlike the theory of degradation toward chaos, this sequence of events is believed to be approaching increasing harmony and organization.

The first formulations of the progressive model of time appeared in the seventeenth century, yet it was not until the nineteenth century that the model came into prominence. Fully two centuries of philosophical development and a host of social and economic changes occurred between the introduction and general acceptance of the idea of progress.

One might wonder why the theory of progress is such a recent phenomenon. Why didn't the idea take hold in the Greek classical period or during the Middle Ages? The answer involves several related economic, social, and philosophical issues. During the Greek era, it was virtually impossible to advocate any notion that society could be changed for the better. The reason was the clear separation, imagined by the Greeks, between the earthly and celestial kingdoms. Human affairs, by their very nature, were considered to be imperfect reflections of the sacred activities of the gods. That is unequivocally expressed in the writings of Plato, who considered human society to be fragile and flawed.

Even Plato's so-called utopian writings reflect his prejudices against the possibility of human progress. His republic was conceived as having occurred during an enlightened period of history that had long since passed. Furthermore, it was unstable; it decayed to less enlightened, despotic forms of government. That vision coincides with Plato's notion of cyclical time. Society, in Plato's viewpoint, begins with a golden age, passes through a series of epochs of degradation, and culminates in a period of decay. Afterward, a new society with a corresponding new golden age, develops from the ashes of the old. This cycle illustrates the insubstantial nature of earthly civilization. Reality can be found only in the world of the gods. An impenetrable barrier separates the earthly and heavenly spheres.

It was considered to be a sin to attempt to perfect human society through scientific advances. That was seen as a sort of theft of heavenly secrets, as can be seen in the myth of the demigod Prometheus. Prometheus brought the secret of fire from the gods down to earth, and for that he was condemned to eternal punishment. Any attempt to penetrate celestial mysteries presumably would condemn any curious person to a similar fate.

Clearly, this antipathy toward scientific inquiry played a significant part in discouraging notions of progress among the Greeks. The Greeks could not imagine significant changes in their society that were due to technological or scientific innovation. Therefore, it was natural for them to think of the world as static by imagining time as cyclical. Cyclical time provided a convenient way of describing a society without progress.

That is not to say that Greek society eschewed science completely. The preeminent scientist of the Greek period was Aristotle, one of Plato's students. Aristotle involved himself in a number of scientific endeavors, breaking, in some ways, the taboo against the organized study of nature. Still, he was not a believer in human progress. As we have discussed, Aristotle, like Plato, supported the predominant circular time idea of the Greeks.

Ironically, Aristotle's writings, revolutionary for their time, came to play a reactionary role in the years after the downfall of Greek civilization. During the Middle Ages, Aristotelian thought exerted a crushing influence on any attempts to break free from its grasp. Consequently, the thousand-year period before the seventeenth century contained little in the way of scientific experimentation, which greatly contributed to downplaying the idea of human development.

Another factor which hindered the rise of the idea of progress in the Middle Ages was the influence of St. Augustine. Although Augustine advocated a form of linear time, he envisioned a catastrophic end of civilization in which the world will be dominated by the supremely evil Antichrist. Therefore, within the framework of time itself, according to Augustine, there is no room for the concept of progress. Only after time itself has ended will paradise be restored.

During the Middle Ages, religious philosophy was dominated by the Christian orthodox interpretation of St. Augustine and secular philosophy and science were devoted to analyzing the works of Aristotle. Deviations from those bodies of thought were viewed as heresies. For that reason, as the great philosopher of history J. B. Bury argues, the idea of progress was suppressed until the Renaissance.[1] One could hardly argue for a viewpoint in which humankind can pull itself up by its own bootstraps when the prevailing orthodoxy claimed that humans are corrupt and evil by nature. That was especially true when the orthodox viewpoint was liberally enforced by death sentences for heretics!

During the Renaissance, the toleration of dissent gradually increased. Creativity in art, music, and literature flourished, and scientific activity increased. As the power of the orthodoxies of the Middle Ages waned, new philosophical ideas could be brought to the surface. In that atmosphere, the idea of progress first appeared.

The seeds of the idea were nourished by the intense interest in exploration and mercantile activity during the sixteenth and seventeenth centuries. The old order, namely the Church and the nobility, depended upon a fairly stable social structure to survive. The mercantile class, on the other hand, needed rapid change and expansion. Naturally, they found the ideas of expansion, development, and progress to be appealing.

One of the first philosophers of the progressive view of human society was an Englishman, Sir Francis Bacon.[2] Born in London in the year 1561, Bacon was an extremely versatile man whose philosophical genius was matched only by his enormous political success. At age 22, he was elected to Parliament. In a short time, Bacon rose to the venerable position of Lord Chancellor of England while writing extremely influential essays on the nature of scientific inquiry. This combination of a cogent writing style and political renown helped him to overcome the scientific inertia of the Middle Ages.

Bacon immodestly viewed himself as another Aristotle and set out to surpass the latter's scholarly approach. His major work on scientific organization, *Novum Organum*, published in 1620, presented a radically new approach to the investigation of nature. Advocating the abandonment of the dogmatic echoing of Aristotlean conclusions, Bacon called for a thorough analysis of natural phenomenon through the accumulation of data and the search for patterns. He advocated the formation of national and international scientific organizations, which would be funded through state support.

Bacon was not a believer in philosophical inquiry for its own sake. The purpose of knowledge, he thought, is the betterment of human-kind—to provide the greatest good for all. As knowledge accumulates, human society will progress. Although Armageddon would surely come in a few centuries, the lot of humanity need not deteriorate before then, he thought.

Bacon was a religious Christian and did not intend to dispute the authority of the Gospels. Nevertheless, his attitude toward the final days of man on earth was quite radical for its time. Bacon did not question the Christian notions of the second coming and the end of time, but he challenged the idea that society must regress into a state of decay before the Armageddon. Bacon imagined a period of growth and prosperity, a time in which humankind gained control over nature before the end of life on earth would occur. Those views presented a counterpoint to the traditional Augustinian approach, because Augustine had insisted that society could not advance on its own accord.

Bacon also specifically refuted the Greek idea of the eternal return. He denounced the idea that fortune must be followed by misfortune as a concept that had caused needless despair. The only reason why progress had not been continuous was that superstition and prejudice had delayed human achievement. Once humankind had been set on the right track by enlightened thinkers, progress would be irreversible. Bacon's last work, written just before his death in 1626, was a utopian fantasy called *New Atlantis*. In it he described a society that was ruled by scientists and governed for the material benefit of all of its citizens. Humanity had successfully exploited nature to provide comfort for all.

Bacon's essays had a powerful effect upon the perception of time and nature by society. For the first time, the notion that nature could be tamed by humankind was advanced; it displaced the idea that human destiny must be determined by uncontrollable natural forces. Thus, an optimistic approach to time gradually seeped into public awareness.

Books advocating a progressive viewpoint began to proliferate during the seventeenth and eighteenth centuries. In 1627, George Hakewell, an English theologian, published a manuscript directly challenging the notion that nature was decaying and humankind was degenerating. In he argued that Christianity had led to social progress; humankind had continually advanced since pagan times. That advance would continue until the end time, when the world would be suddenly consumed by fire.

Note that Hakewell argued for an upward slope model of time. He held that society will progress until time itself is brought to an end by the hand of God. There is no connection, in Hakewell's view, between the progress of humankind and the end of the world.

By the end of the seventeenth century, optimism had become quite attractive to European thinkers. The philosopher Leibnitz maintained that the universe must be heading toward a perfect order. The reason is the perfection of its creator: How could an omnipotent God create anything less than a universe striving for perfection?

The French satirist Voltaire mocked that viewpoint in *Candide*.[3] Nevertheless, he felt that human progress was inevitable. Unlike Hakewell and Leibnitz, Voltaire argued that progressive changes occurred *in spite of* Christianity! Without the corrupt church and government officials, Voltaire envisioned a society that would improve immensely.

Turgot, Voltaire's contemporary and fellow countryman, went a step beyond his predecessors by developing a social theory of humanity's progress. In his manuscript, "Discourses on Universal History," he searched for the causes of progress. By that time (1750), divine providence was no longer viewed by most thinkers as a factor which played a significant role in human history. That being the case, Turgot looked for what we would now call psychological and sociological determinants. Those factors include passions, ambitions, and natural talents, as well as ethnic and geographical conditions.

Turgot did not assume that higher moral values such as reason and justice played leading roles in shaping human destiny. In fact, he argued that wars, revolutions, and exploitation were important factors in reshaping the political scene. He concluded that, in spite of human folly and greed, there had been an irreversible tendency for society to progress.

Turgot divided human intellectual development into three periods. In Turgot's scheme, the first stage is one in which natural phenomena are explained by supernatural causes. This stage is followed by a period in which the legends of the gods are discredited but science has not yet been developed. Therefore, abstractions are used to explain natural events. The final period is the age of science and reason.

By the beginning of the nineteenth century, the idea of progress had become an integral part of most philosophical systems. The French and American revolutions had set the stage for dramatic changes in social structure throughout the western world. Meanwhile, industrial growth drastically altered the economic structure of Europe and America. Scientific discoveries served to limit the perceived role of divine intervention in nature. All these changes heralded a new spirit of innovation and growth.

Religious philosophers searched for ways to incorporate spiritual notions into a progressive, scientifically valid approach to history. In Christian theology, the Augustinian notion of social regression was largely discarded. Instead, a progressive viewpoint emerged, one in which human progress was seen as reflecting a divine perfecting principle. Leibnitzian philosophy played an influential role in that movement.

Undoubtedly, the leading Christian philosopher of history during the nineteenth century was the German thinker Hegel, who viewed history as the unfolding of the spirit of God on earth.[4] Each stage of political development coincided with a corresponding spiritual development. This progress began in the Orient, traveled west to India, to Greece, and to Rome and finally to the Prussian state of Hegel's day. As unbelievable as it sounds now, Hegel viewed the nineteenth-century Prussian monarchy as the culmination of human political and spiritual development. No further progress was possible or necessary. God's spirit had finally been realized on earth in the form of Germanic culture and morality.

As it turned out, the spiritual philosophers of progress did not play a substantial direct role in the formation of nineteenth-century political culture. Economics, and not religion, had become the dominant influence in society. For that reason, Hegelian thought was important only in its adaptation to material-based philosophies. Materialism had become the basis of most nineteenth-century theories of progress.

Progress and Materialism

During the Industrial Revolution, the notion of progress came to be associated with increased prosperity for all. It was felt that science and technology would lead to an abundance of material goods for all citizens. It was only a matter of time before comfort and happiness would be a birthright.

In marked contrast to that hope, the economic situation in much of western Europe was bleak. Thousands of people left their farms for the cities, only to find themselves in impoverished surroundings. In eighteenth- and nineteenth-century Britain, child labor was quite common. As a "progressive" measure to clear the streets of beggars, workhouses were set up for the poor. Factories were filled with paupers who worked over 12 hours a day for meager wages. Class differences grew, and the bulk of the population engaged in lives of utter desperation. The downward slope model of time would have seemed more appropriate for someone observing the situation. How could anyone speak of human progress, in light of such misery for the majority of people?

It takes a special gift to find order when chaos is all around. Only people with a certain amount of foresight and imagination can make that leap of faith. Adam Smith, the father of modern economics, was one of the rare breed. Smith was born in Scotland, in 1723. At age 28, he assumed the chair of logic at the University of Glasgow. He soon won renown for his studies of moral philosophy. Particularly interested in the question of selfishness, Smith analyzed how human self-interest can be overcome by objectivity and impartiality. His book on the subject, *The Theory of Moral Sentiments* drew much attention throughout Europe.

In 1776, the year of the Declaration of Independence of the American Colonies, Smith produced his masterpiece, *The Wealth of Nations*. This book has been called "the outpouring not only of a great mind, but of a whole epoch." In his highly influential work, Smith reveals a method by which economic order can emerge from chaos. He provides a means by which a so-called invisible hand can guide society toward prosperity and mutual benefit in spite of the pettiness and short-sightedness of individual motives. His aim, in the broadest sense, is to outline how the wealth of a nation can continue to grow indefinitely, given the nature of the human condition.

Smith argues that unrestricted competition and free trade will lead naturally to the accumulation of wealth. Individual self-interest guides the manufacturers of goods to produce what the public wants. Furthermore, the prices for those items are kept reasonable by the simple fact that, if the prices are too high, another manufacturer can step in and produce the items more cheaply. Since specialization leads to more efficiency and more profit, Smith sees a general tendency toward the differentiation of labor. Finally, he predicts that the mechanism, if unhindered, will lead to increased production, wealth, and social prosperity without the need for any guiding structure.

Smith found a method by which randomness can lead to order on an economic level. Progress arises in spite of apparent social turmoil. As we shall see, there is an analogy between this sort of formulation and that of model physical theories of turbulence. In each case, large-scale structure emerges out of local chaos. As Robert Heilbroner puts it: *"The Wealth of Nations* had its own laws of motion....Adam Smith's world went slowly, quite willingly, and more or less inevitably to Valhalla."[5]

Many flaws in Smith's analysis have been found. Modern capitalism is far too complicated to be described by his arguments. The Great Depression and numerous ups and downs in the business cycle have tellingly convinced many of the need for more sophisticated approaches. Even in Smith's day, arguments to dispute the notion that economic progress is irreversible were found.

One of the outstanding problems in Smith's scheme was found by Parson Thomas Malthus in 1798. With the dire forecast embodied in his treatise "An Essay on the Principle of Population as It Affects the Future Improvement of Society," Malthus single-handedly dashed the hopes of many new converts to the notion of progress. His simple thesis served as a substantial argument for the downward-arrow picture of time, at least on a human scale, predating the law of entropy and the gloomy predictions of Ehrlich and Rifkin.

What is this devastating argument against sustained progress? Malthus shows, in a clear manner, that economic growth cannot be sustained forever. The problem lies with the growth of the human race, requiring the development of new resources to sustain this ever-increasing population. Quite simply, the population grows in a geometric manner (1,2,4,8,16,...), whereas the development of food sources and land increases arithmetically (1,2,3,4,...). Eventually the need for food and land cannot be satisfied, since the population grows at a much faster rate.

This situation cannot be addressed by a laissez-faire philosophy of economics. The only possible solution is that of population growth planning. Although some might argue that the "invisible hand" proposed by Smith might lead to restraint, it is not clear that it would be automatic. Certainly the idea of mass starvation as a check to growth is not a pleasant prospect. Advanced action to control growth seems to be a much more humane solution.

During the nineteenth century, a number of visionaries proposed highly organized models of the ideal society. Strict planning replaced laissez-faire as the favored scheme for progress. Unlike Smith, many felt that laissez-faire would lead to economic anarchy, social chaos, and mass deprivation.

The Utopian socialists, thinkers on the cutting edge of progressive thought, developed schemes to bring about social equality, increased prosperity, and controlled population growth. Because they advocated the optimistic point of view that the world would inevitably get better, they played an important role in furthering the idea of the upward slope model of time.

Robert Owen, for instance, felt that by changing the social structure, paradise on earth could be attained. This Welsh industrialist earned an international reputation for designing a model community in New Lanark, near Glasgow, where workers could live in harmony. Buying several old mills, Owen aspired to be the exemplary industrialist. His workers lived in large, clean homes and worked shortened hours under favorable conditions.

Continuing his experiment, Owen set out to establish "villages of co-operation" around the world. In these villages, 800 to 1200 people would live communally, sharing kitchens, living rooms, and dining halls. Within these communities, decent jobs could be found for all, effectively ending poverty. Children could be educated properly because there would be no need for child labor.

Owen's social experiment met with little success. His American village of cooperation, New Harmony, was short-lived and inconsequential. Nevertheless, Owen's ideas had a strong influence on the British labor movement. Workers began to support the idea that they could take destiny into their own hands and create a worker's paradise. Perhaps scientific innovation could result in shorter working hours, safer conditions, and cooperative ownership of the means of production. The first trade unions and consumer cooperatives inspired by the vision of Owen, were formed in Britain at this time. In 1833, the Grand National Consolidated Trades Union was formed as an attempt to organize the English working class. At the same time, the Rochdale pioneers created the first organization to represent consumer interests. These societies promoted the notion of worker and consumer cooperatives.

Meanwhile, utopian socialists in France painted their own pictures of the perfect society. Count Claude Henri de Saint-Simon spoke of the appalling waste associated with the leisure class. He argued for a principle by which all citizens should be expected to produce roughly the same amount of work. Government should be organized along economic lines in order to promote social equality. In his work, he sketched an outline of how this society would come about, but he did not provide many details.

Charles Fourier, on the other hand, had a meticulous program for the reorganization of society. Unfortunately, most of his ideas bordered on the fantastic. For instance, he envisioned that the seas would turn into lemonade and that six moons would appear in the sky when the future paradise would be realized. Nevertheless, Fourier had a strong influence on the utopian movement with his more conventional ideas. Society was progressing, he thought, to higher and higher levels of organization. Eventually all people would be organized into social structures, called phalanxes, which would provide for prosperity and equality.

Not all the progressive French social philosophers of the nineteenth century had such abstract notions of social harmony. Alexis de Tocqueville found his ideal in the newly founded United States of America. He saw in America a society in which class differences were largely eliminated and democracy was flourishing. Europe and the rest of the world would move in the same direction.

Utopian literature was extremely popular during the nineteenth century. From Owen and Fourier to H. G. Wells and Edward Bellamy, writers looked beyond the harshness and grime of life in the industrial age toward a sparkling new future of order and affluence enjoyed by all. Few thought that what was the contemporary state of affairs could last forever.

The utopian movement of the last century differed dramatically from the golden age concept of the Greeks. Plato had proposed that each cycle of history begins with a golden age, which is constructed by the gods. Modern utopian literature imagines a perfect society created by humankind itself. It can therefore be associated with an upward slope, rather than a cyclical, model of time. As Bury has remarked, in contrast to the myth of a golden age, utopia is "...man's effort to work out imaginatively what happens, or what might happen, when the primal longings embodied in the myth confront the principle of reality. In this effort man no longer merely dreams of a Divine state in some remote time; he assumes the role of creator himself." On the other hand, "a characteristic of the Golden Age...is that it exists outside history, usually before history begins."[6]

Both the utopian socialists and the laissez-faire capitalists were searching for an organizing principle—a means by which social conditions would ensure a gradual movement toward progress. The socialists found it to be a desire for cooperation and social harmony; the followers of Adam Smith argued equally vehemently that the answer is to be found in unshackled competition. The two groups held radically different notions of human nature. Of course, there were other viewpoints. The political philosophy of the German economist Karl Marx combined the egalitarian vision of the socialists with the pragmatism of the advocates of laissez-faire capitalism.

In nineteenth-century Germany, the greatest influence on philosophical discussion was the work of Hegel. It is not surprising that Hegelian thought had an enormous influence on the writings of German political thinkers, notably Karl Marx. While at university in Bonn and Berlin, Marx pondered the Hegelian notion of constant change, which Hegel had called dialectics. Marx was fascinated by the concept of finding a purpose to history. However, unlike Hegel, he was an atheist and a socialist. Therefore, his notion of the meaning of historical change was radically different from that of his philosophical predecessor.

Marx proposed the concept of dialectic materialism: dialectic change in the Hegelian mode with a material rather than spiritual basis. In the Marxist scheme, in other words, economics rather than spirituality and the realization of God's will on earth provides the driving force of history. According to dialectic materialism, the means of production of any society supplies the framework for the structure of that society. For instance, slave labor provided the production mechanism of Roman civilization, whereas industrial technology supplied the material basis for capitalism and European parliamentary democracy. In order to make any predictions about history, one must understand the underlying economic mechanisms which bring about change.

Hegel felt that the Prussian monarchy was the culmination of history; Marx begged to differ. Marx thought that the German government and its associated social order were far from stable. In his work, he detailed the underlying mechanisms which he believed would lead to revolutionary change around the world.

Marx possessed the ability to deal equally well with arguments over obscure philosophical abstractions and clear calls for revolutionary action. In some of his writings, he makes sweeping statements intended to rally the workers. In other works, he seems obsessed with minor details of interest only to specialists. For that reason, it is possible to draw many contradictory conclusions from Marx's work. As a result, Marxism has evolved into thousands of different forms. In many places, the writings of Marx have become official dogma resulting in a situation similar to that of the church in the Middle Ages. In fact, the Marxist and religious movements have many parallels.

Both Marxism and Christianity support a linear approach to time with a timeless paradise following the end of history. In Christianity, this period follows the apocalypse and the second coming. Marx says the final stage of history is communism.

According to the Marxist interpretation of history, called historical materialism, there are five major stages of human development. The first stage is called primitive communism: the state of affairs before the coming of imperial conquest and war. The second period of development was represented by Roman civilization, in which slave labor was the underlying economic mechanism. According to dialectics, internal contradictions brought about the demise of the Roman empire. These contradictions led to the overthrow of one ruling clique by another clique. Thus the feudal lords gained the upper hand as the Roman empire fell.

The third period of history is that which we call the Middle Ages. Marx labeled it the feudal period; it ended when the rising capitalist class began to consolidate its power. As the economic basis of society shifted from agriculture to industry, the feudal lords lost power to the rising middle class.

The fourth period of history is that in which the world is dominated by the capitalists. Many contemporaries of Marx assumed that capitalism was a natural state of affairs and would thus last forever. Marx found this assumption to be amusing, considering the demise of feudalism to be proof of the dynamical character of history. Marx writes in his *Poverty of Philosophy*, "The economists regard bourgeois institutions as natural and based on eternal laws, and feudal institutions as artificial. Thus there has been history, but there is no longer any."[7]

Marx argues that capitalist institutions are also fragile. Historical change would occur because of the organization of the working class. Eventually the working class would assume control over society. This period would be called the "dictatorship of the proletariat." After this period of worker rule, all class distinctions would be eliminated. History would draw to a close, because there would be no more contradictions in the social structure. Marx calls this the period of world communism.

It is interesting that Marx argues against the point of view that capitalist society would last forever but at the same time preaches of an eternal period of communism. One might wonder why Marx assumes that the dynamic element of history would disappear at this point. Christopher Dawson, the philosopher of history, explains this apparent contradiction in the following manner:

> Clearly [this inconsistency] is due to the victory of the Marxian apocalyptic over the Marxian philosophy. It is the essence of apocalyptic to look to the end of history and it can never be content with an endless movement of cyclical change. And the apocalyptic hope meant more to Marx than all his rational theories."[8]

Today the world is dominated by two superpowers with two competing philosophies of progress. Both the Soviet and American governments have adopted ideologies which argue for an upward movement in history: the Soviet model based on Marx and the American model based on Smith. Unfortunately, both societies have made the mistake of Marx in arrogantly assuming that current social structures will continue into the future. Thus, we find that American propaganda suggests that the American economic and political system is the goal of history, and the Soviets suggest that their own society should be the model for other governments (though "glasnost" has somewhat tempered this assertion). The fact that both superpowers subscribe so dogmatically to the progressive view of history provides an indication of the strength of this point of view in the world today.

The Theory of Evolution

It can be argued that the downward slope model of time didn't really enter the realm of science until the second law of thermodynamics was proposed by Clausius. The law of entropy lent a certain amount of credibility and ominous certainty to the idea that the universe is running down. Similarly, the upward slope model of time was confined to the philosophical arena until Darwin's revolutionary proposal that the human race is the end product of a long evolutionary chain of events. With the arrival of the theory of evolution, the progressive, optimistic view of time was transformed from speculation into science.

Darwin's proposal didn't just come out of the blue. A series of speculative works on geology and other natural sciences set the stage for the acceptance of an evolutionary model of species development. In many ways, Darwin was a genius, but it cannot be said that he was ahead of his time. The paleontologist Stephen Jay Gould suggests that the linear arrow of time, which he refers to as "deep time," appeared in geology as early as 1690 with the publication of the Reverend Thomas Burnet's *The Sacred Theory of the Earth*. Burnet's influence is looked upon unfavorably by present-day geologists because he based his natural speculation solely on the Bible and did not use the scientific method. Nevertheless, Gould finds a progressive undercurrent in Burnet's proposals.

In his work, Burnet begins with the assumption that the Bible is a completely accurate account of the earth's history. To this he adds a radical proposal: that physics and natural causes can explain all of the Biblical events. For instance, he explains the great flood by detailing a method by which the earth's crust cracks and underground water is brought to the surface as a deluge. During the deluge, most of earth's geological formations are fashioned by erosion. Thus, Burnet speculates that earth was much more attractive before the flood. Postdiluvian earth is a "hideous ruin" by comparison.

Burnet tries to reconcile cyclical and linear approaches to geological time. On the one hand, he argues against the Greek time approach of absolute repetition. On the other hand, he speculates that physical forces will return the earth to its original state. This is similar to Poincaré's idea of recurrence, Nietzsche's eternal return, and the modern oscillating universe approach. In Burnet's approach, however, all the physical elements of Earth will be restored to their original positions through a second conflagration as predicted in the Bible. Burnet emphasizes that the second global event will be different from the first, thus distancing himself from the cyclical approach. This future catastrophe will be of fire rather than water. In spite of its different nature, the second catastrophe will serve as a mirror image of the first. Following the great fire, paradise will be restored on earth in much the same manner as it was taken away by the great flood.

As Gould points out, there is an interesting interplay between cyclical and linear time in this prediction. He views Burnet's work as:

> ...the finest expression ever published of the tension between two complementary views of time—the ancient contrast of time's arrow and time's cycle...the Sacred Theory is a playground for Burnet's struggle to combine the metaphors into a unified view of history that would capture the salient features of each—the narrative power of the arrow, and the immanent regularity of the cycle.[9]

With Burnet's work one can see the fragile nature of the attempts by scientists to correlate their work with the Biblical description of events. The attempts were soon to be abandoned. By the middle of the nineteenth century, most scientists treated the Bible as a metaphorical document having no relationship with scientific theory. Geological dating techniques led most naturalists to believe that the earth is much older than the age implied by the sequence of events in the Old Testament. How can we explain such a dramatic turn of events in the field of geology?

A highly influential geological work appeared in 1795. In his *Theory of the Earth*, the Scottish geologist James Hutton argues against the catastrophic theory of geologic change. By asserting that the earth's topography was formed by slow evolutionary processes, he pictures a global history that is quite different from the Biblical version. Hutton postulates that the forces that form mountains and rivers have always operated in the same way. The earth today looks very much like the earth thousands of years ago. This is called the principle of uniformitarianism. Since the geologic features that we see around us have been formed through erosion and other slow processes, the earth must be millions of years old. Hutton speculates that the age of the earth might even be infinite. There is no reason to believe that the earth will end in a catastrophe.

Hutton's approach to geologic time is not linear. As in traditional circular theories of time, Hutton's world looks pretty much the same now as it did 1 million years ago. However, by speculating that the earth is so old, he lends credibility to the idea that natural evolution could, in fact, have taken place. Before Hutton, natural science was careful to support, at least in principle, the idea that the earth is only several thousand years old.

Expanding upon Hutton's work, the geologist Charles Lyell, in 1830, published the first volume of his book *Principles of Geology*. Lyell describes in great detail his studies of geologic formation. He concludes that Hutton's premise is correct: Topographic features are the result of slow processes such as volcanic action and erosion. There have been no global catastrophes such as the great flood. Geologic forces continue to act today in exactly the same manner as they did in the past.

The work of Hutton and Lyell paved the way for a new description of natural phenomena. With the creationist theory of the earth eliminated as a possibility, biologists needed to explain the existence of the various life forms that are found all over the globe. The idea of divine intervention was no longer acceptable. A theory that could account for the variety of all the species was needed.

Numerous theories by which higher forms of life could evolve from lower forms were proposed. It was thought that evolution was the logical alternative to creationism. However, it wasn't until Charles Darwin published his *Origin of Species* in 1859 that a comprehensive explanation of the evolution of life was presented. Darwin conducted an extensive program of research aboard the H.M.S. Beagle. He looked at life in its various forms and noted the abundance and variety of species. He wondered how certain species flourish and others die out. He developed an explanation of evolution: the theory of natural selection.

Natural selection could, in Darwin's view, account for the variety of species and for the evolution of one species into another. In this theory, life evolves through a process that involves several steps. The first step of the process involves variation. Some mechanism unknown by Darwin, produces different offspring from the same species. Today we believe that this process involves mutation. Radiation and other natural processes alter the DNA in the reproductive cells of a plant or animal. Since the DNA serves as a genetic code to determine the makeup of the next generation of a species, the offspring of the plant or animal turn out to be mutated: they possess traits different from those of the parents.

The next step of the process is natural selection. Because of variation, different species and varieties possess unique characteristics. Some of these properties enable the creatures to survive better in their environment; others cause the species to have less chance of survival. The characteristics that lead to the best chance of survival and propagation are the ones that are clearly favored. This mechanism leads to an evolutionary process by which different species emerge while others become extinct.

In his later work, *Descent of Man*, published in 1871, Darwin details how humankind has evolved from a relative of the apes via an evolutionary process. Details of this evolution can be found in fossil records and skeletal remains. Darwin's theory had an enormous influence on both the scientific community and the general public. In some sense it provided a material basis for the progressive model of time. Through evolution, animals, humans, societies, and perhaps even the universe as a whole could transform into higher and higher states of development. The social sciences welcomed Darwinism as a validation of the idea of progress. In many ways, Darwin was the most influential scientist of the 1800s. As Boltzmann remarked in 1886, during an address to a meeting of the German Imperial Academy of Science:

> If we regard the apparatus of experimental natural science as tools for obtaining practical gain, we can certainly not deny its success. Unimagined results have been achieved, things that the fancy of our forebears dreamt in their fairy tales...Nevertheless I think that it is not these achievements that will put their stamp on our century: if you ask me for my innermost conviction whether it will one day be called the century of iron, or steam, or electricity, I answer without qualms that it will be named the century of the mechanical view of nature, of Darwin.[10]

The first sociologist to promote the theory of Darwin was his fellow Englishman Herbert Spencer, who had anticipated Darwin's works with his own theory of evolution published in 1857. When *Origin of Species* came out, Spencer immediately recognized it as a work of universal importance. He incorporated it into his theory of the evolution of human society, and his works were highly influential. Darwin had a great deal of respect for him and once wrote that he considered Spencer to be "by far the greatest living philosopher in England; perhaps equal to any who ever lived."

In 1858, Spencer began to plan a series of books surveying the fields of biology, psychology, sociology, and morals from an evolutionary perspective.[11] He had already written an article, "The Nebular Hypothesis," describing a scheme by which the stars, planets, and other objects in the galaxy have evolved from gaseous clouds. He imagined that the universe has evolved from chaos into orderly structures. His *First Principles*, the first volume in his planned series, appeared in 1862. It was soon followed by *Principles of Biology* and *Principles of Psychology*, both following the evolutionary approach. The first volume of *Principles of Sociology* appeared in 1876.

Spencer's ideas on sociology can be described by the term "survival of the fittest," which he coined. He argued that the weaker elements in a society die out through war, poverty, and natural disaster. Only the strongest people and societies survive. In this manner, an evolutionary process has taken place, and in it wars and disasters, although quite horrible, have served an important purpose. Civilization has reached higher and higher forms through the weeding out of undesirables.

Later sociologists and economists used the example of Spencer in proposing an incorporation of Darwinian evolutionary theory into the social sciences. The idea of survival of the fittest was appealing to followers of both Smith and Marx. For laissez-faire Smithian economists, the use of the expression "survival of the fittest" echoed what they had always believed. Those who are clever can produce a better product and dominate the market, they argued. People who cannot create a usable item or provide a useful service do not find work or make a profit. The best industries survive; the worst become extinct. The "invisible hand" of economics is merely another name for natural selection.

Some more conservative economists and sociologists took this a step further. A philosophy called social Darwinism came into being at the end of the nineteenth century. People who supported this view held that certain economic classes, races, and social groups are inferior to others. Through a struggle for survival, the strongest of these groups will come into power. Social Darwinism was used to justify the worst excesses of industrial society, for it meant that the poor deserved their lot because of their inferiority. Even worse, it implied that efforts to help the weak and poor were counterprogressive.

Marxists interpreted Darwin in a radically different manner. Marx thought that his own theories of class struggle were similar to Darwin's model. In fact, he even wanted to dedicate a portion of *Das Kapital* to Darwin, feeling that his own work paralleled *The Origin of Species*. The class struggle in Marxism is similar to Darwin's notion of the struggle for survival: that, through a battle between the workers and the capitalists, the weaker shall perish and the stronger shall survive.

Marx probably didn't literally mean that the capitalists or "reactionaries" of any type should be exterminated, yet once again a theory was taken to its extreme. Dogmatically following his own interpretation of Marx, Joseph Stalin slaughtered millions of peasants in the name of evolution and progress. One can see how one's interpretation of the meaning of time and history can be quite a dangerous thing!

One of the most important supporters, and critics, of the idea of evolutionary social progress was the British writer H. G. Wells. Wells wrote several utopian novels in which evolution led to higher states of political growth. Yet as soon as he painstakingly built these edifices of social harmony, he proceeded with wit and cynicism to knock them down. In the *Time Machine*, for instance, the time traveler travels to a future earth which seems like paradise.[12] Flowers have evolved into more beautiful, delicate forms. Human society appears to be harmonious. Yet the protagonist soon realizes that humanity has begun to *devolve* into more primitive forms. What was once the upper class has become a frail, helpless subspecies of humanity. What had once been the proletariat, on the other hand, has been turned into a society of subterranean brutes.

As the time traveler ventures further into the future, he finds that higher life forms are replaced by more and more primitive ones. Eventually, at a time far in the future, all life has died out. The planets are slowly orbiting closer to the sun. It seems that the world will soon come to an end. The time traveler realizes to his horror that entropy, not progress, will have the upper hand in the distant future.

The sentiments of Wells reflect the debate over the validity of imagining evolution as an indefinite process. Wells was aware of the theories of Darwin and Spencer, but he was also aware of the viewpoint of Clausius. If one accepts all these theories, one must ponder the dilemma caused by the sharp clash between two radically different views of time. One view is represented by evolution, the other by entropy. The former corresponds to the upward slope model of time, the latter to the downward slope model. The problem comes with reconciling the two points of view.

The law of increasing entropy implies that disorganization is increasing; the law of evolution seems to suggest otherwise. Life appears to be evolving toward forms that are more and more complex. For example, no one would argue that an amoeba and a human being are equally complex. A universe cascading toward a heat death and one producing structures of greater and greater complexity present a profound paradox. Is the universe dying out, or is it just beginning to come to life?

One possible solution to this problem is to postulate that life is an exception to the second law of thermodynamics. As the philosopher Henri Bergson wrote, "Life ascends the slope that matter descends."[13] Some have argued that the second law can be applied only to systems that are close to equilibrium. (An equilibrium state is one in which no significant change is possible without an outside force.) It does seem entirely possible that the second law is inapplicable to nonequilibrium systems, since Clausius formulated the law by using the assumption that the thermodynamic states that he considered were close to equilibrium. Moreover, he considered only closed systems, ones which exchange energy, but not matter, with their surroundings. The process of life would not be the sort considered by Clausius. Living beings always exist in a state that is far from equilibrium; the equilibrium state of life is death. Also, they are open systems that continually exchange matter with their environment. Therefore, the entropy law might not be applicable to living processes—a second law escape clause.

Arthur Koestler makes a strong, almost pedantic, argument against the inevitability of the heat death in *The Ghost in the Machine*:

The gospel of flat-earth science was Clausius' famous second law of thermodynamics. It asserted that the universe is running down, like a clockwork affected by metal fatigue, [toward] the Cosmic Heat Death....Only in recent times did science begin to recover from the hypnotic effect of this nightmare, and to realize that the second law applies only in the special case of so-called closed systems....That this Law did not apply to living matter, and was in a sense reversed in living matter, was indeed hard to accept by an orthodoxy still convinced that all phenomena of life could ultimately be reduced to the laws of physics.[14]

Koestler argues for an "optimistic" arrow of time to replace the pessimistic point of view held by most physicists:

The overall view of biological and mental evolution reveals the working of creative forces all along the line towards an eventual realization of the potentials of living matter and living mind—a universal tendency towards the spontaneous development of greater heterogeneity and complexity.

The forces that Koestler appears to be battling against are those that have tried to incorporate living matter into thermodynamic models. There has been a strong resistance in the physics community to the possibility of any exemption from the law of entropy. It is generally argued that living beings operate by a process of energy degradation, i.e., they convert orderly sources of energy (food) into disorderly waste energy. The thought processes engaged in by humans involve a clear creation of order from disorder. However, it is claimed that the amount of entropy decrease due to this information increase is more than offset by entropy increase stemming from metabolism. Thus, the net effect is still an overall *increase* in entropy.

One might, however, note that the amount of entropy decrease due to creativity is not something that can be measured in a precise manner. Consciousness is something that is extremely hard to quantify because it is still little understood. That makes it difficult to reach any clear conclusions about the application of the second law to *conscious* organisms. As Du Nouy points out:

In first approximation it can be said that all vital phenomena are dependent on the laws of energetics, and in particular, on the Carnot-Clausius law. They consequently contribute to the increase in entropy of the system of which living beings are a part, just as all the other phenomena resorting to chemistry and physics.

On the other hand,...the experimenter must never forget that the living being forms a complete organism. It has a personality, and the biological phenomenon as a whole is not simply due to the summation of elementary chemical phenomena, but to the order in which these phenomena occur in time and space. The order appears to be the expression of a pre-determined purpose.[15]

The classic work on the application of thermodynamics to living organisms is Erwin Schrödinger's *What is Life?* Schrödinger explains that life is significantly different from other physical processes. It does obey the law of entropy—but is capable of maintaining significant reservoirs of order amid the general chaos of its surroundings. As Schrödinger puts this: "It feeds on negative entropy."[16] Life does reach a state of maximum entropy, namely, death, but it takes an extremely long time to reach that state in comparison to the rate by which most physical processes reach equilibrium.

He then argues that this anomaly can be explained in terms of physics. Life is organized by chromosomes—aperiodic structures that are largely shielded from entropy-increasing processes. The amount of damage caused by the environment usually doesn't significantly alter the makeup of these genetic blueprints. Consequently, by reproducing via genes, life can maintain states of complexity for long periods of time. Only radiation damage to the chromosomes themselves can stop this orderly reproduction.

Here we find something quite remarkable. The genetic chemicals of life (which we now refer to as DNA) provide a protection against the ravages of the environment. They help to slow down entropy-increasing processes. Thus, there are two essential ingredients for the sustenance of life. One of these is material—the availability of food and other raw materials as a source of negative entropy. The other is organization—the presence of DNA as a genetic blueprint for reproduction.

DNA is one of the most complex substances known to humankind. It consists of a set of four different chemicals, called bases, arranged as rungs on a helical ladder. These chemicals act as letters in a genetic language and dictate, through an indirect mechanism, what sort of proteins should be produced by a cell. Moreover, DNA has the power to replicate—to produce a copy of itself. It is truly miraculous that such an intricately structured substance could have evolved.

Many scientists have argued that the evolution of DNA is an extremely unlikely event and could not have happened during the relatively short history of the earth. Statistically that seems to be true. It is hard to imagine that such a long and complex molecule could form randomly—even considering the fact that the environment of the early earth was far more volatile, leading to far more mutations, than today. Some scientists, notably Fred Hoyle, have suggested that DNA has an extraterrestrial origin. That would allow for the possibility of DNA evolving over a longer period of time. There has not been much evidence to support that radical proposal.

Another possible solution to this problem invokes the anthropic principle. Even if the set of conditions leading to life is extremely unlikely, it is a fact that we are here today to consider this issue. There are billions of other possible worlds in which nobody can make any statements regarding evolution — since the unlikely event of the development of intelligent life has not occurred there. Our own existence suggests that the improbable has occurred, that life has evolved on our planet.

Of course, that argument can be refuted if life is discovered elsewhere in the universe, which would seem to suggest the *inevitability* of life, rather than its unlikelihood. It would be quite surprising, for example, if intelligent creatures arrived from Betelguese carrying copies of the *Times* in their briefcases! We would have to reconsider our notion of evolution as a chance happening.

A number of philosophers have invoked what is called a "design argument" with regard to evolution. Simply put, a design argument is a statement that the complexity of an object suggests the existence of a creator. Here the delicate structure of DNA causes some to assume that it was molded by the hands of a divine being. Instead of the old way of thinking about God — that God created all forms of life at once — the new evolutionary theological approach is to consider evolution as having been "programmed" by God.

One of the leading advocates of that position was the paleontologist and Jesuit priest Teilhard de Chardin.[17] Teilhard's unconventional method of combining Catholicism with Darwinian theory earned him the scorn of the Jesuit movement. Publication of his philosophical works was forbidden during his lifetime. Upon his death in 1955, his writings finally appeared. In part because of their controversial nature, they caused quite a sensation.

In a manner similar to Hegel, Teilhard viewed human progress as the unfolding of God's will on earth, but he took that a step further. He proposed that all of evolution and all of human history has been the realization of Christ on earth. As human society becomes more complex, it is forming closer and closer links of communication, which Teilhard called the noosphere. Eventually, the noosphere will contract to a point, the omega point. The intelligence of the universe will become one. He equated this organized intelligence with the spirit of Christ.

According to Teilhard, there exist two realms: the physical and spiritual. Each realm possesses its own kind of energy, respectively the tangential and radial energies. The tangential energy experiences the sort of degradation predicted by Clausius. If all processes in the universe

were powered by that type of energy, then the universe would decay
until it reached a heat death.

The power that drives the ever-increasing complexity of the universe
is radial energy; Teilhard thought it is what caused the simplest proto-
zoa to evolve into increasingly complicated forms of life. It also has provided the drive for human progress and the quest for human unity. In
fact, the ultimate goal of civilization is the organization of humanity on
a grand scale. Only then can the omega point be reached. Teilhard
called the whole process Noogenesis.

John Barrow, the British cosmologist, has pointed out the similarities
between Teilhard's picture of the omega point and the modern cosmological description of the end of time for a closed universe (one that
collapses).[18] Unlike the downward slope closed-universe model, in
which the end state of the cosmos is completely quiescent (found, for
example, in Hawking's approach), Teilhard's proposed end of time is
far more optimistic. Teilhard believed that the spirit of intelligence
could outlast even the heat death. The omega point would then be a
time of universal harmony. Teilhard held that:

> ...there is a progress, within us and around us, a continual height-
> ening of consciousness in the universe.
> For a century and a half the science of physics was dominated by
> the idea of the dissipation of energy and the disintegration of mat-
> ter. Being called upon by biology to consider the effects of synthesis,
> it is beginning to perceive that, parallel with the phenomenon of
> corpuscular disintegration, the universe historically displays a sec-
> ond process as generalized and fundamental as the first: I mean that
> of the gradual concentration of its physicochemical elements in nu-
> clei of increasing complexity, each succeeding stage of material con-
> centration and differentiation being accompanied by a more ad-
> vanced form of spontaneity and spiritual energy.
> The outflowing flood of Entropy equalled and offset by the rising
> tide of a Noogenesis![19]

In the past few decades, more conventional methods, well grounded
in experimental research, have been applied to try to reconcile the two
arrows of time represented by entropy and evolution. New theories of
self-organization that have emerged may form the seed of a complete
explanation of the origin of life and other complex phenomena, viewing these systems as islands of order in a chaotic decaying universe.

Self-organization

Twentieth-century physics traditionally has encompassed two models of
the direction of time. The first model is the circular road viewpoint,

which first entered the realm of physics in the time of Newton. As we have seen, that picture of time can be successfully applied to the microscopic world of particles, where the order of events is of little importance. The second, downward slope, approach is applied to situations related to macroscopic systems, as in problems involving heat or matter exchange. A great deal of theoretical work has gone into explaining why ignorable time can be utilized in some cases whereas unidirectional, entropic time must be used in others.

The third model of time's path, which may be called the upward slope approach, has found applications in the world of physics only recently. Traditionally, the optimistic arrow of time has been neglected by physicists. One might think of several possible reasons for this lack of interest in progressive time:

1. The idea of progress is relatively new. As we have detailed, the progressive view of the world was suppressed until the late Renaissance. It was not until the nineteenth century that it came into prominence. Even then, it did not appear in the sciences until the time of Spencer and Darwin.

2. Physics had become increasingly isolated from the rest of the scientific community. The hallmark of the Industrial Revolution had been specialization. This had been especially true in the world of the sciences. Until quite recently, mainstream physicists had little interest in developments in biology, chemistry, psychology, meteorology, and so on. In some of these other fields, discoveries that have been made (such as evolution) have not permeated the physics community. Fortunately, that is changing; now there is far more cooperation among the sciences.

3. The full understanding of complexity was impossible before the computer age. The development of the computer has resulted in a radically new approach to physics. Before the coming of the computer, the motto of physics was "simplify." The result was great progress in understanding two sorts of extreme cases.

The first traditional approach to understanding a certain physical phenomenon is to consider only one or two objects with the greatest possible symmetry present. For example, the law of gravity was discovered by Newton's consideration of the orbit of the moon around the earth (two objects), of the motion of the earth around the sun (two objects), and of the dropping of single objects, like apples, toward the earth (two objects). By confining himself to simple examples, Newton soon discovered the inverse-square law form of gravity. The same approach has been used to understand electrical and magnetic phenomena.

The second traditional method of producing physical models is to consider a very large set of objects and to look for statistical relationships. That is the method by which the laws of thermodynamics have been truly understood by physicists. The science of statistical mechanics has been developed to understand the properties of large sets of molecules. For example, since a typical container of gas contains trillions on trillions of molecules, it is possible to look at the *average* behavior of the set. By doing so, one is simplifying enormously. For example, if the gas is far from equilibrium, it is very difficult to make statistical approximations. Thus, nonequilibrium thermodynamics is not very well understood.

With one or two objects, we see a Newtonian reversibility. If we consider an infinite set of objects in equilibrium, we see thermodynamic irreversibility and an approach to heat death. What about the middle ground?

Until the coming of the computer, which has become cheap and available only in the last two decades, this middle ground was unexplored territory. It was assumed, however, that it looked something like one of the two extremes. That is like assuming that the United States is heavily forested and densely populated after exploring just the east and west coasts. Clearly, the properties of the state of Nebraska would not be found by such an approach. It is only now that these vast hinterlands of physics are being surveyed with the help of computer simulations.

Computers have proved invaluable to contemporary physics. With these tools, systems of nontrivial complexity can be fully explored for the first time. Experimental conditions can be duplicated on a computer, with the result that highly sophisticated models can be developed. It is only with these models that we can see a different sort of pattern of temporal flow.

Another way in which computers have been of great use is in the field of artificial intelligence. By examining the problem of how to accomplish certain tasks by the use of a computation device, we have developed more insight into the structure of information accumulation and processing. We can begin to tackle the formidable question of the workings of the human brain and the meaning of consciousness. Thus, the whole question of the evolution of intelligence can be dealt with via physics for the first time. As we saw in the preceding section, this problem has profound consequences for the meaning of time.

4. Progressive time cannot be understood by an algorithmic scheme. During the past few decades, we have witnessed an interest in a holistic approach to science. Traditionally, physicists have followed an algorithmic method of problem solving: a rigid method involving step-by-step calculations was utilized to obtain an answer. Normally, that is associated with a completely deterministic approach: given a set of initial con-

ditions, future values of some function can be found to match the experimental data. The holistic point of view is one that is more interested in spotting large-scale patterns, rather than systematically computing the values of the details, of some organized structure. The holistic idea is closely linked with the idea of intelligence.

We can see the importance of holism by imagining trying to apply the traditional scientific method to the world of art. For instance, the image of the *Mona Lisa* has appeared in thousands of guises. One can find this famous picture on computer-generated posters, on television, and in cartoons. Grotesque distortions of the *Mona Lisa* can be found in all sorts of media; as an example, one might consider Duchamp's depiction of *Mona Lisa* with a moustache. Suppose we wished to discover the essential feature of this image: What makes a *Mona Lisa* a *Mona Lisa*? An algorithmic method could not possibly yield an answer to that question. One would become absorbed in the particulars of the physical or chemical makeup of the molecular structure of a given copy of the portrait, which would, of course, vary substantially from one medium to another. Only a holistic approach, one which emphasized pattern recognition, could be employed to yield an answer.

During the 1960s and 1970s, there was an enormous increase in the public awareness of holism. Holistic approaches to health care, for instance, became increasingly popular. It is probably true that the popular interest in viewing objects in terms of totalities provided a ready group of scientists eager to apply the principle to nature. James Gleick has suggested that the radical, nonconformist spirit of the 1960s played a part in providing the drive for a novel approach to physics. Certainly an interest in a holistic lifestyle could lead one to contemplate holistic theories of physics.

If we combine the idea of evolutionary progress with a holistic approach to nature, coupled with an interdisciplinary interest in computer simulation, we can readily see how a new scientific paradigm has appeared. Yet in the true holistic spirit, the whole is much greater than the sum of its parts. The new physical scheme that has developed is the principle of self-organization. This principle has emerged from a pooling of thoughts from all fields of science. Self-organizational schemes provide a fundamentally progressive outlook regarding the nature of time. Through these approaches, a full understanding of the emergence of order from chaos, and hence of the upward slope model of time, has become conceivable.

To see how order can stem from disorganization, one must first understand the true nature of physical chaos. In the past, chaos was thought of as the manifestation of the lack of knowledge of some physical system stemming from the fact that the system has a great number of components. If all the elements forming a certain system could be known exactly, an exact physical model could be constructed.

For example, during the 1950s it was thought that weather prediction was destined to become an exact science. If one could take enough measurements of the atmosphere at any one time, one could construct a computer-generated model of the atmosphere's future states. The mathematician and meteorologist Edwin Lorenz decided to put the doctrine to the test. He constructed an ideal model of the weather by using a few equations of minimal complexity. By plugging in values of the velocities of air currents and the values of other parameters, such as temperature, pressure, and the rate of the earth's rotation, he thought that he could make weather forecasts.

Instead, Lorenz found that his model was extremely sensitive to initial conditions. A slight change in any of the parameters would lead to an exponentially increasing deviation in the forecast. That proves to be catastrophic for the model, for it is impossible to know any natural parameter with complete precision. There is always some uncertainty in its value.

What Lorenz discovered is sometimes called the butterfly effect: the property of some sets of equations that any small change in the initial conditions would lead to an ever-growing lack of certainty in the future values of those quantities. Such equations are not deterministic; knowledge of the past does not guarantee knowledge of the future unless the past is known perfectly well. Any small fluctuation can lead to random results. Thus, even the flapping of the wings of a butterfly can alter the outcome of the weather in an unpredictable manner!

Since the early sixties many other examples of physical systems modeled by sets of equations which display sensitive dependence on initial conditions have been found. In the current scientific terminology, these systems are said to exhibit the property of "deterministic chaos." The hallmark of deterministic chaos is that simple equations can lead to astonishing complexity.

If that were the complete story, we could scarcely have cause for hope that the complexity of the universe can be described by simple equations. In fact, chaos theory seems to model randomness and turbulence. It is hard to imagine how such a method could be used to describe organization.

Interestingly, many chaotic systems exhibit a great deal of regularity. For instance, models similar to the weather simulation of Lorenz, which are called strange attractors, have been developed. These objects display a curious mixture of chaos and pattern formation. The equations for these attractors act, for all intents and purposes, like random-number generators. Yet, in spite of the apparent randomness, strange patterns can be found in the results. If one plots the results of the equations on a graph, all the points lie on bandlike structures. If one looks closely at them, the bands exhibit substructures of thinner bands. In fact, there is an astonishing degree of regularity to the picture that is formed. The geometric design that emerges is a structure, called a fractal, of enormous complexity.

The moral of chaos theory is that even simple sets of equations can produce seemingly random results, and that even seemingly random results can exhibit intricate patterns. The interesting thing about the patterns is that they are surprisingly universal and independent of the specific equations chosen. That has important implications for the nature of the scientific search for the origin of organized structures. It may be impossible to find out which equations can be utilized to model these structures, for a whole class of equations might lead to the same sort of organization. What is of much more interest is to search for the universal features of mathematical models. That is what we mean by a *holistic* approach to problem solving. Instead of trying to look for a specific equation to model a certain set of data, it is of more significance to search for the general method by which patterns can emerge from randomness.

What, then, do we mean by self-organization? Self-organization is the unpredictable appearance of structure amid essentially random results. The emergence of this complexity is completely independent of the natural laws which generate the results. It arises from little known universal properties of mathematics.

For example, Mitchell Feigenbaum has found that many equations can exhibit both simple and chaotic results depending on the value of a certain parameter (adjustable constant) in the equations. By adjusting the parameter (much as one would turn a dial to tune a radio), one can go from regular to chaotic sets of results. The interesting thing is that the rate in which one goes from simplicity to chaos is largely independent of the equations themselves. Feigenbaum has found universal constants that measure this rate.[20]

Theorists studying chaotic systems come from many different disciplines. Rather than focusing on the particular properties of specific equations, they are interested in universal features of mathematical models; they hope to apply the universality to natural phenomena. They hope to find the properties by looking for patterns in computer-generated maps of the results of the equations. That is the same holistic approach that our brains utilize to distinguish the *Mona Lisa* from other paintings or from random sets of dots.

The theory of self-organization can lead to an optimistic arrow of time in spite of the second law of thermodynamics. If patterns can emerge from randomness, then it is possible to imagine a mechanism by which organized structures, such as the DNA helix and the human brain, can stem from seemingly random molecular encounters. As Gleick puts this, "Somehow, after all, as the universe ebbs toward its final equilibrium in the featureless heat bath of maximum entropy, it manages to create interesting structures."[21]

Paul Davies, the British cosmologist, views the theory of self-organization as the basis of a new paradigm in physics:

> For three centuries science has been dominated by the Newtonian and thermodynamic paradigms, which present the universe either as a sterile machine, or in a state of generation and decay. Now there is the new paradigm of the creative universe, which recognizes the progressive, innovative character of physical processes. The new paradigm emphasizes the collective, cooperative, and organizational aspects of nature; its perspective is synthetic and holistic rather than analytic and reductionist.[22]

Self-organization theory and the idea of chaos have been applied to a host of natural systems. Scientists have been interested in learning how order can arise from randomness in a variety of circumstances. Chaos theory has led to possible models of the formation of the rings of Saturn to solve the question of how rocks engaged in random motion have settled into intricate patterns. It has been applied to models of fluid turbulence, resulting in the discovery that order has been found amid turbulent flows of water. Ecologists have been interested in the question of why certain populations of ecosystems grow, others die out, and still others tend to oscillate.

One of the most exciting areas of research has been in the area of artificial intelligence. Research workers have developed theories of learning called neural network models. Here scientists have been interested in how the acquisition of knowledge can take place so rapidly. Learning cannot take place by a linear, algorithm process; it would simply take too long. Thought patterns must arise in a different manner, probably through a holistic method similar to the formation of complexity in chaotic systems. It is hoped that someday we will be able to fully comprehend the mechanisms of the human brain and understand how they evolved.

A question that is of fundamental interest is how life developed from lifeless chemicals. One of the most fundamental characteristics of living matter is its ability to reproduce itself. In particular, how can one chemical organize an exact copy of itself from the materials at hand? It is unclear how self-reproducing life evolved, for without reproduction in the first place it is difficult to imagine evolution at all.

In order to better understand the idea of reproduction, the mathematician John von Neumann set out in 1950 to create mechanical models of self-duplication. He called these models automata. The automata would, given the proper instructions, be able to construct exact copies of themselves from the surrounding materials. The copies would automatically contain the instructions on how to build more copies until the machines would proliferate.

After satisfying himself that the construction of such a machine was physically possible, von Neumann decided to model automata mathematically. By constructing an automaton on a sort of checkerboard, he developed the first of what are now known as cellular automata. A cellular automaton consists of a lattice (grid) of cells in which each cell can assume one of a given set of values. Based on simple rules which change its values, each cell is updated periodically. The rules depend on the values of the cell and its neighbors.

To model a self-reproducing machine, von Neumann constructed a numerical pattern on a grid. He then devised rules by which each lattice site can be updated. After a sufficient amount of time, he found that the original configuration can reproduce an exact copy of itself by using the simple rules.

Since von Neumann's time, cellular automata have been used to model a variety of physical, chemical, and biological systems. The interesting thing about the automata is the tendency for patterns to form. There is nothing encoded in the automata rules which tell the cells how to create the patterns. The patterns arise spontaneously and provide an intriguing example of self-organizational behavior.

In 1970, an automaton model which generated a lot of interest among computer programmers was constructed by John Conway, a mathematician at the University of Cambridge. The model, called the Game of Life, fascinated observers with its seemingly endless catalog of fascinating patterns spontaneously generated by the automaton rules. The game's intent is to model the evolution of the populations of species with factors such as overcrowding and starvation playing regulatory roles by keeping the growth rate down.

To play this game, one starts with a grid or checkerboard in which each square has the value 0 or 1. Cells with value 0 are considered to be dead; cells with value 1 are considered alive. When a living cell is surrounded by too many other living cells, it dies because of overpopulation. On the other hand, a living cell needs at least a few neighbors to survive and to reproduce.

All those rules for deaths and reproduction are applied at once throughout the checkerboard during a single step, called an iteration. After each iteration the pattern of living cells changes. Conway found, to his surprise, that self-organization of a most intriguing form can occur in the automata.[23] By starting with random configurations of 0s and 1s (dead matter and living organisms), he noticed an evolution toward interesting organized structures. Organisms would appear to "eat" other organisms in a struggle for survival. Moving structures called gliders would move across the screen. Some patterns would die out; others would exhibit growth; and still others would exhibit periodicity, repeating themselves after a certain fixed amount of time.

Cellular automaton models show us how order can stem from chaos.[24] For most scientists, these models are just toys. They simulate, but don't duplicate, real life. The true picture of the universe is far more complicated. There are, however, physicists who believe that cellular automata are more than just simulations. Edward Fredkin, formerly at MIT, is the leading advocate of the view that the universe itself is an automaton. Fredkin feels that it is just a matter of time before someone stumbles upon the correct set of automaton rules for the universe. That would constitute a formula by which all the structures that we see today have evolved from a completely random pool of information. Information structures, not particles, form the basis of everything we know. Particles are simply patterns of non-zero information, much like gliders in the game of life. Andrew Ilachinski, along with the physicist Max Dresden, has developed a variation of this theme. Here the universe is a topological automaton, a lattice in which both the site values *and* the geometry change according to automaton rules. Particles are seen to be disturbances (like waves) in the lattice.[25] It is unclear, in all cellular automaton models of the cosmos, how one can make connections to the physics as we know it. For that reason, little work has been done on developing cosmological cellular automaton schemes.

The interesting thing about some of the models in which order emerges from chaos is that the second law of thermodynamics appears to be violated. Entropy decreases as patterns evolve from random initial formations. In order to explain how the models relate to real thermodynamic systems, one must reconcile their properties with the law of entropy increase. It is of significant interest in the scientific community to develop a means of understanding the relationship between arrowless determinism in the Newtonian fashion, the downward slope of entropy increase, and the upward slope of pattern formation.

Perpetual Motion

An interesting method of relating the two possible directions of time's arrow has been developed by the physicist Ilya Prigogine.[26] Prigogine has detailed how random fluctuation can lead to increasing states of order in some cases and to increasing disorder in others. He has shown how patterns can arise from purely chaotic behavior.

Prigogine's means of describing the formation of order from chaos is in some ways analogous to Darwin's theory of natural selection. In Darwinian evolution, species experience random fluctuations that are due to their environment. If these fluctuations are not too great, the species tends to remain stable. It continues to reproduce copies of itself that do not vary significantly. In some cases, however, the variations are very great. In that instance, one of two things may happen. If the new variation is more favorably suited to the environment than the original species, the former will propagate in greater numbers than the latter. The species is said to have evolved to a higher form. If the variation is unfavorably suited, the mutated version of the species will die out.

The key to Prigogine's description of order arising from randomness is the idea of a series of states arranged in order of complexity. At any given time, a physical, biological, or chemical system is in one of the states and the situation is fairly stable. Because a small genetic mutation may not significantly affect a species, in general, a small fluctuation would not bring the system out of that state. There may, however, come a time when a large fluctuation (which may be caused by an outside factor or may just be due to the mechanical properties of the system) causes the configuration to find itself close to another one of the stable states. That state may be more complex than the original state. In addition, it may be more environmentally favorable for the system to be in the new state. Hence, complexity may evolve spontaneously from chaos through an evolutionary process. In short, what happens is that random fluctuations can lead a system to reach new "islands of stability," new configurations that are stable but are more complex than the original states.

A complete understanding of the concept of stability requires a grasp of the intricacies of the modern physical theory of *quantum mechanics*. Quantum mechanics provides a theory of how processes can either remain stable or decay depending upon certain conditions. Generally, physical systems have a certain probability of decaying into states of lower energy. Between these times of decay, however, many structures display remarkable durability.

Under certain circumstances, quantum mechanics permits a sort of perpetual motion to take place. Some configurations do not decay at all. In those cases, there is the possibility that certain processes will continue forever until an outside agent brings them to a halt. Under those circumstances, time ceases to have a directional arrow. Such situations are of special interest because they seem to suggest that the clockwork regularity of the Newtonian universe is possible under certain circumstances.

A good example is the case of superconductivity. In normal conduction, electrons are drawn through a wire. Passing through impurities, the electrons give up some of their energy as heat. The process conforms to the law of entropy increase. Under some circumstances, however, perfect conduction can take place. Normally this occurs for very low temperatures (although recent experiments indicate that there may be the possibility of room-temperature superconductivity). Here quantum mechanics has required that the electrons pass by any impurities without giving up energy. Therefore, entropy does not increase. Electrons traveling in a superconducting loop may continue to circle the loop forever. Therefore, they experience a kind of perpetual motion or timelessness.

It appears that the question of the circularity or linearity of time may be answered only when quantum mechanics is fully understood. Only through quantum theory can we understand the related issues of decay, stability, and fluctuation. Therefore, the nature of the form of time is still a problem that remains to be solved.

We have shown that the progressive model of time, like the circular and "pessimistic" models, has its advocates in the scientific community. That is for good reason, because the progressive world view has been the dominant one over the last two centuries. With the tremendous growth in the theory of self-organization, the upward slope model of time has developed a stronger theoretical foundation. Perhaps the structural uniformity associated with certain systems in which quantum mechanical effects produce large-scale regularities will yield even more of a basis for this point of view.

In the next chapter we shall further explore the quantum mechanical notion of time, especially with regard to the issue of simultaneity. We have completed our study of the *direction* of time and now wish to understand the relationship of events within the fabric of time itself.

References

1. J. B. Bury, *The Idea of Progress*, Macmillan & Co., London, 1921.
2. Will Durant, *The Story of Philosophy*, Simon and Schuster, New York, 1961.

3. Voltaire, *Candide,* New American Library, New York, 1961.

4. G. W. F. Hegel, *Lectures on the Philosophy of World History,* Cambridge University Press, Cambridge, 1975.

5. Robert L. Heilbroner, *The Worldly Philosophers,* Simon and Schuster, New York, 1953.

6. J. B. Bury, *The Idea of Progress,* op. cit.

7. Jacques Barzun, *Darwin, Marx, Wagner,* Doubleday, Garden City, N.Y., 1958.

8. Christopher Dawson, *Dynamics of World History,* Sheed and Ward, New York, 1956.

9. Stephen J. Gould, *Time's Arrow, Time's Cycle,* Harvard University Press, Cambridge, Mass., 1987.

10. Ludwig Boltzmann, *Theoretical Physics and Philosophical Problems,* D. Reidel Publishing Company, Boston, 1974.

11. Herbert Spencer, *The Evolution of Society,* University of Chicago Press, Chicago, 1967.

12. H. G. Wells, "The Time Machine," in *Three Prophetic Novels,* selected by E. F. Bleiber, Dover, New York, 1960.

13. Henri Bergson, *Creative Evolution,* Macmillan Publishing Co., London, 1964.

14. Arthur Koestler, *The Ghost in the Machine,* Macmillan Publishing Co., New York, 1967.

15. P. LeCompte Du Novy, *Biological Time,* Macmillan Publishing Co., New York, 1937.

16. Erwin Schrödinger, *What is Life? The Physical Aspect of the Living Cell,* Cambridge University Press, Cambridge, 1945.

17. Pierre Teilhard de Chardin, *The Future of Man,* Harper and Row, New York, 1959.

18. John Barrow and Frank Tipler, *The Anthropic Cosmological Principle,* Oxford University Press, New York, 1986.

19. Pierre Teilhard de Chardin, *The Future of Man,* Harper and Row, New York, 1959.

20. Mitchell Feigenbaum, "Qualitative Universality for a Class of Nonlinear Transformations," *Jour. Stat. Phys.,* **19** (1978), pp. 25–29.

21. James Gleick, *Chaos,* Viking Press, New York, 1987.

22. Paul Davies, *The Cosmic Blueprint,* Simon and Schuster, New York, 1988.

23. J. H. Conway, unpublished.

24. Stephen Wolfram, *Theory and Applications of Cellular Automata,* World Scientific, Singapore, 1986.

25. Paul Halpern, "Sticks and Stones: A Guide to Structurally Dynamic Cellular Automata," *Am. Jour. Phys.* **57** (1989).

26. I. Prigogine, *Order out of Chaos — Man's New Dialogue with Nature*, Bantam Books, New York, 1984.

4
Sending Signals

Time and Communication

Quite independent of the question of the *direction* of time is that of the *structure* of time. In the former case, one asks "Where is it all heading?," with "it" referring to one's own life, the collective existence of the human race, or the duration of the cosmos as a whole. In the latter case, one confines one's thoughts to the present moment and inquires into the nature of the succession and duration of events. The appropriate question would seem to be "How are we getting to where we are heading, and are we all proceeding at the same rate?"

One might draw an analogy with riders in a horse race. The riders might at times be interested in the nature of the horse track. They might ask, "Is the track circular, or does it go in a straight line?" Or they might ask, "Is the track sloping uphill or downhill?" These are the sorts of questions we have dealt with so far in this book.

On the other hand, a rider might want to know "Am I ahead of or behind the other riders?" "Are there any shortcuts in this race?" "Can I send a message to one of the other riders?" These are questions that are related to the ideas of succession and communication. The latter half of this book will deal with such issues.

Communication is a concept that is closely related to the notion of causality. Without any sort of communication, every part of the universe would comprise its own island. There would be no relationship between the parts that make up the whole. No event could be said to influence any other event.

Communication can be said to take place in one of two ways. First, two objects can touch each other. In that way, one object can perceive something about the second object. The two objects must be at the same place at the same time for that to happen. This can be called local communication.

The second possibility is that one object sends a signal to a second object. The two objects occupy different positions and do not come into direct contact. Here one can distinguish two separate events. The first event is the sending of the signal. We refer to it as the *cause*. The second is the reception of the signal and the resulting change in the second object. It is called the *effect*.

Let us consider several examples of the process. The first is that of the electromagnetic interaction between two charges. In that case, one charge creates what is called an electromagnetic field. The field propagates through space in the form of light radiation. The radiation then affects the motion of the second charge, carrying electromagnetic energy. The second charge "feels" a new force on it, which causes it to accelerate in agreement with Newton's laws of motion.

In this example, it is clear what is cause and what is effect. Moreover, one can thereby see how communication is closely related to causality. Here the cause lies in the emission of light, whereas the effect lies in its reception. The carrier of information is light radiation in the form of wave packets. This method of relaying information is used in wireless communication such as radio transmission.

There are other forms of communication that are slower than light wave transmission. Information can also be carried by mechanical means, via so-called mechanical waves. Water waves, seismic waves, and traveling atmospheric disturbances are examples of the phenomenon. Information carried by mechanical pressure waves in a material medium such as air or water is called sound. Sound waves are much slower than light waves because they are dependent on material pulsation rather than vacuum transmission. Sound waves do not involve an actual exchange of material; instead there is an exchange of information via a succession of material disturbances.

The mechanism involved in transmitting sound waves is rather like the conveyance of traffic information along a motorway or highway. Suppose traffic is rather heavy along a stretch of highway. Suddenly a car comes to an abrupt stop. Drivers behind the first car react as fast as they can to bring their own cars to a halt. The information travels slowly backward through the queue. In that way the presence of cars slowing down in front of each driver sends a signal to all the other cars along the road to also slow down. When the first car starts up again, the signal travels the same way. Nothing material is actually exchanged between the first and last cars, but a signal is transmitted anyway through the motion of the cars.

A radio signal would be a far more efficient means of causing all the cars to stop. Suppose every driver simultaneously heard the message, "Stop your car immediately!" All the cars would stop almost at once. There would be an almost instantaneous transmission of the information.

Someone viewing the scene from above would be surprised by the effect. The person would have to conclude that, because of the near simultaneity of the action of the cars, a faster means than a pulse in the line of cars would have to have been employed. The viewer would deduce that either the drivers arranged to stop at a preset time or a radio signal was sent out.

Most exchanges of information in nature occur between two objects at different places. Naturally, there is a built-in time delay between cause and effect. No one has yet found a faster means of transmission than via electromagnetic waves (i.e., radio waves, visible light waves, x-rays, etc.). In other words, no message can be sent at a rate exceeding the speed of light in a vacuum.

Before the famous Michelson-Morley experiment of the late nineteenth century, it was thought that all waves were material waves. Light, it was supposed, traveled through a material called ether. Ether could not be detected; nevertheless, it filled all of space. Just as the speed of a sailboat depends on whether travel is with or against the wind, it was postulated that the speed of light would depend on the direction of the flow of ether relative to the earth or, rather, the direction of the earth's motion relative to the fixed ether.

Michelson and Morley attempted to measure the effect of the earth's velocity, relative to the ether, on the speed of light. To their surprise, they found that the speed of light is constant; it is independent of the direction of etherial flow. They were reluctantly forced to abandon the ether model; they concluded that, in fact, light propagates in a vacuum and is not a material wave of any sort.

Their experiment was quite elegant in its simplicity. In their laboratory in Cleveland, Ohio, they set up an apparatus consisting of a series of mirrors. A beam of light was split into two parts. Each part of the beam was reflected by mirrors and sent back to the source. The arrangement was such that half the light traveled in the direction corresponding to that of the earth's motion and the other half traveled in the perpendicular direction. If there were an ether background, there would be a time difference between the two beams similar to that between the motions of two sailboats setting off in different directions relative to the wind. The team found no time difference at all. The experiment has been repeated thousands of times with the same null results.

Special Relativity

The principle of the constancy of the speed of light, which the Michelson-Morley experiment seemed to imply, clashed with a fundamental notion of classical physics: the principle of relativity. According to the principle, all the laws of physics should be independent of the velocity of the observer. That statement can be found implicitly in Newton's laws of motion. As we recall, those laws state that a body at rest will remain at rest and a body moving at a constant speed will remain at the same speed unless it is acted on by an unbalanced force. There is nothing in nature to distinguish a state of rest from a state of constant motion. In both cases, the body is said to be in equilibrium, a condition in which all forces are balanced. Only if there is *acceleration*, meaning a change in the speed or direction of motion, can we infer that there is any change in the condition of the object.

For example, suppose one were to play a game of pool on board a train. Assume that the train is moving in a straight line with constant speed. While playing the game, there would be nothing to indicate that the train is moving. The pool balls would bounce off the sides of the table in exactly the same way they would if the train were not moving. Only if the train accelerated would there be a difference in the motion.

A problem arises if one attempts to reconcile the classical principle of relativity with the notion that the speed of light is constant. We can see that by continuing with our train analogy. According to relativity, there should be no experiment which can distinguish between a rest frame and a moving nonaccelerated frame (with "frame" meaning "vantage point"). However, imagine the following scenario:

Consider a train moving parallel to a beam of light at the same speed as the light. From the rest frame, one can measure the speed of light and come up with its standard value. On the other hand, from the moving frame of the train, the light appears to be standing still. That follows from the fact that if two objects move at the same speed, they appear to be at rest with respect to each other. Thus, one obtains two different values of the speed of light: one from the rest frame and one from the moving frame. That contradicts the principle of the constancy of the speed of light.

To reconcile the principle of relativity with the principle that the speed of light is constant, Albert Einstein developed his theory of special relativity.[1] We have already encountered Einstein's theory of general relativity, which presents a new way of looking at gravitation. Special relativity deals with the relationship between two observers moving at constant velocity with respect to each other. It provides a framework for comparing measurements taken in two different frames.

In special relativity, the idea of absolute space and absolute time are abandoned. Unlike the Newtonian framework, spatial displacement is considered to be *dependent* on the velocity of the observer. The same applies for temporal duration. In other words, an observer traveling in a moving frame would perceive time to flow at a different rate from that of one in a static frame with respect to the earth. The observers would also measure spatial distances differently.

We have already pointed out that Einstein advocated an amalgamation of space and time called space-time. Here we see how the Newtonian distinction between space and time is blurred by special relativity. If an observer sitting on a frame that is fixed with respect to the earth observes a set of events that occurs on a moving frame, it appears that the time interval between the events is smaller than the observer would expect. For example, clocks would appear to run slower in the moving frame. On the other hand, spatial distances on the moving frame, as viewed from the rest frame, appear to be shorter. A yardstick located on the moving frame would appear to be shorter than 36 in.

In short, a frame moving with a constant velocity relative to another frame undergoes a shift in *space-time*. Spatial distances are shorter, and temporal durations are longer. The only distance that stays the same is the distance in space-time. Here space and time are almost placed on equal footing. The only difference between a spatial and temporal displacement is that a spacelike distance is taken to be *real* and a timelike distance is taken to be *imaginary*. We use the term "imaginary" in a precise mathematical sense; an imaginary number is the square root of a negative real number. It is not necessary to grasp this distinction mathematically; it is sufficient for our understanding of the issue to simply realize that space and time are not treated in exactly the same way in relativity but there is, nevertheless, a close relationship between the two concepts.

Let us explore some of the implications of the special theory of relativity. First of all, relativity implies a certain kinship between space and time. Some philosophers and scientists, particularly those attracted to the circular notion of time as presented in Chap. 1, have used this relationship as an argument for the "geometrization" or "spatialization" of time: the concept that the universe must be viewed as a static, four-dimensional whole. Others have pointed out that, since time and space are not quite treated on an equal footing in relativity, the idea of flowing or linear time still makes sense. One might still imagine the universe as an evolving three-dimensional object.

These questions relate to the issue that we have discussed in the last few chapters, namely, the debate over whether or not time has a direction. The idea of considering space and time as one entity is anathema to one who advocates an evolutionary point of view. How could evolution or regression take place if the universe is a completely determined space-time object?

Another fundamental issue is the question of the uniformity of the flow of time. How can we measure the flow of time (assuming that time does flow) in an objective way if every object in the universe contains its own clock which runs differently depending upon the relative speed of the object? The answer is we can't. No one can claim that any particular clock shows the objectively correct time.

Thus, when we speak about the flow of time of the universe as a whole, we must now take into account that the universe is a patchwork of an infinite number of clocks that are running at different rates. None of these clocks measure the proper time of the universe. We can speak of such a time, but we cannot determine its local value directly.

Does all this mean that, in fact, we cannot speak of time having a flow at all? That is not necessarily the case, since Einstein's theory carefully considers the relationship between cause and effect. If event A follows event B in one frame, it must follow event B in all moving frames. The time between the two events may not be fixed, but the order is. The guarantor of the preservation of the direction of causality is the idea that the speed of light is the speed limit of the universe. If it weren't for that fact, the course of events could be seen differently in different time frames. Therefore, even though the rate of flow of time might be different for various observers, the direction must be kept fixed. (There is a proposal for particles that travel faster than the speed of light, and hence violate causality, but it is highly speculative.)

The fact that no object's velocity can exceed the speed of light has strong implications for the process of communication. Signals must travel at a finite rate. Therefore, there cannot be a causal connection between events taking place at a sufficiently large distance away from each other. The distance is that for which light cannot travel in a given period of time. If, for example, two events take place at times which are a second apart and the distance between the events is greater than a light second (the distance that light can travel in one second), then the events cannot possibly be causally connected. In other words, one event cannot be the cause of the other, because information cannot proceed from one event to the other in the amount of time supposed.

One can draw diagrams (called Minkowski diagrams) of the circumstances in which two events can and cannot be causally connected.

A Minkowski diagram consists of two axes labeled space (horizontal) and time (vertical). Drawn on this diagram are two crossed lines intersecting at the junction of the spatial and temporal axes and representing the paths that light takes from a given event. In a vacuum, light always takes a straight path at a constant speed; hence its route can be traced out in a Minkowski diagram as a set of crossed straight lines of constant slope (3×10^8 m/second). Thus, speeds of paths above and between the crossed lines would be less than the speed of light and speeds of paths below and next to the lines would be above the speed of light.

Now let's see how the diagram can be used to determine whether or not two events can be causally connected and in communication with each other. Consider one event to be at the intersection of the axes (the origin). If the second event is plotted above and in between the crossed lines, the connection between the events is said to be *timelike*. That means it is quite possible for a signal to travel between the two events because such a signal would have a speed less than the speed of light. On the other hand, if the second event is plotted below and next to the lines, only a *spacelike* connection is possible. That means that no physical signal could travel between the events because such a signal would have to be faster than light. Thus, the Minkowski diagram provides a visual way of distinguishing between slower-than-light communication, which is possible, and faster-than-light communication, which is believed to be impossible.

Let us return to our motorway analogy to fully explore the fact that communication has a speed limit, namely, that of light. Imagine a set of police cars spread out over the length of a highway. Each police car is equipped with a radio transmitter and a receiver. Suppose the driver of the leading car decides that all of the cars should immediately stop. This driver relays a message to the next driver, who immediately passes the message on. Eventually each driver receives a transmitted signal telling his or her car to stop. Suppose that each car possesses a perfect braking device that stops the car automatically upon reception of the radio message. Also, imagine that the messages are automatically transferred from one car to the next.

What would happen, once the message is sent out, is that all the police cars would appear to stop at once. However, a careful measurement of the time in which each car stopped would reveal that is not the case. There must be a time lag between the cars because of the fact that the signal travels at a finite speed, namely, the speed of light.

Now suppose that all the cars *do* stop at once. We must assume that each car has been told *ahead of time* that it has to stop at the prearranged instant. In other words, information cannot travel faster than the speed of light, but simultaneity *can* occur when it is prearranged.

The whole question of simultaneity is of vital importance when one deals with quantum mechanical phenomena. Quantum mechanics, developed in the beginning of this century, has proved to be an excellent theory of atomic and subatomic physics. As we have seen, quantum systems can display a remarkable amount of organization, which can lead to spontaneous pattern formation. In many cases, it appears that synchronous motion occurs, behavior in which information appears to travel faster than the speed of light. Let's see how quantum mechanical theory seems to suggest the possibility of faster-than-light communication.

Quantum Mechanics and Communication

According to the late Richard Feynman, the American physicist who made outstanding contributions to the simplification and popularization of quantum theory, the key to a comprehension of quantum mechanics is the double-slit experiment, "a phenomenon which is impossible, absolutely impossible, to explain in any classical way, and which has in it the heart of quantum mechanics."[2] This famous experiment illustrates the essential feature of quantum mechanics: a blurring of the dichotomies between waves and particles, between message and sender, and between experiment and experimenter.

The double-slit experiment consists of a screen with two holes in it that acts as a barrier. Beyond the screen is a second screen that acts as a detector. In addition, there is a source providing the subatomic particles called electrons.

Note here that we refer to electrons as *particles*. Together with protons and neutrons, electrons are components of atoms, which, in turn, are the constituents of the natural world. Traditionally, particles have been contrasted with waves, with particles forming the material world and waves being disturbances carrying energy through matter or space. Particles were considered to be solid entities such that, if two particles hit a detector at the same point, there would be double the mark of one particle. On the other hand, waves were considered to be somewhat less than solid: two waves impacting at the same place might cancel each other in the phenomenon called interference.

The double-slit experiment removes the distinction between particles and waves. Imagine a stream of electrons emanating from the source and heading toward the two slits. There are clearly two possibilities: the electrons can pass through one slit or the other. Suppose that the impacts of the electrons are recorded on the detector screen. On performing this experiment, one finds that, instead of a purely additive picture expected for ordinary material objects, a wavelike interference pattern emerges. In the center of the detector, no marks are seen, instead of a large set of them. The impacts of the electrons cancel each other. This result is remarkable, considering that electrons are traditionally considered to be particles, not waves. One might expect that outcome for light or sound, but not for particles!

Suppose one were to cover one of the holes. The electrons could now travel through only one slit and not the other. Surprisingly, one would find that there was no interference pattern at all. By altering the apparatus, the electrons' natures seem to have been transformed from waves to particles. If one were now to remove the cover, even if one were to do so while the electron was traveling between the screens, the picture produced by the detector would show a wave pattern again. The information that the second slit is open seems to travel to the electrons instantaneously!

The interference patterns produced on the detector represent a lack of knowledge about which of the two holes the electrons have passed through. The curious thing is that if one constructs a device to determine the slit in which the electrons have entered, the interference pattern disappears again. The interaction between experimenter and apparatus has immediately altered the nature of the electron from wave to particle.

The property that a particle also has a wavelike nature is called the wave/particle duality. It forms the basis of modern quantum theory, yet it remains a deep mystery. The standard explanation of this phenomenon, called the Copenhagen interpretation, is still controversial over a half-century after its proposal!

According to the Copenhagen approach, all particles may be seen as entities known as wave functions. Wave functions provide a probabilistic picture of where an object, such as an electron, is located. In the double-slit experiment, the wave function is sensitive to the fact that there are two possibilities and hence appears on the detector screen as an interference pattern. (Actually, what is seen is a *probability distribution,* which is, in fact, the square of the wave function.)

When a scientist takes a measurement, the wave function is said to "collapse" into one of its values for that particular quantity. As an example, we can consider an electron in the double-slit experiment. Before a measurement of the electron's position is taken, the electron has a probability of being anywhere in a certain range of locations. The electron's probability distribution encompasses both slits, as if part of the electron passed through each slit. After a measurement of the electron's position is made, the wave function collapses. That means it has a fixed value for its position. So if it is determined through which slit the electron passed, the probability distribution of the electron is that of a particle passing through a single slit. Note that here the experimenter has a profound effect on the wave function. The effect is instantaneous:

the collapse of the wave function cannot be said to take any time what-soever.

Einstein was deeply disturbed by the implications of the Copenhagen interpretation. In his famous statement, "God does not play dice with the universe," Einstein revealed his distaste for the introduction of purely probabilistic notions into the framework of atomic physics. Many consider Einstein's statement to be another of his big blunders. (His first big mistake was assuming that the universe is static; see Chap. 1.)

Einstein hoped that by pointing out the paradoxes of quantum theory he could pave the way for a deterministic approach to the subject. Collaborating with Boris Podolsky and Nathan Rosen, Einstein developed an apparent discrepancy in the laws of quantum mechanics. Although this "thought experiment" could be refuted by Niels Bohr, an expanded version of the problem developed by David Bohm serves as an interesting study of the elusive nature of quantum mechanics.

In this thought experiment, sometimes called the spin paradox, a special property of elementary particles is exploited to yield an interesting conclusion. The property, known as spin, represents a component of the so-called angular momentum of subatomic objects. Roughly speaking, spin is the orientation of the particle with respect to an external magnetic field. For an electron, the spin can hold one of two possible values ($\pm \frac{1}{2}$). We can refer to an upward or downward spin, each corresponding to a different value.

Here we imagine a pair of electrons associated with each other in a singlet state. That simply means one of the electrons has a spin axis pointing up and the other has an axis that points down. The singlet state can easily be created in the laboratory. Now we picture the state breaking up. One of the electrons heads in one direction and the other heads in the opposite way. Still, one electron spins up while the other spins down; we just can't tell which is which.

Detectors have been placed on both sides of the apparatus to record the spins of the electrons. The detector on the left-hand side is so oriented that it picks up signals from electrons of just one type of spin, let's say "up." According to quantum theory, the other detector must pick up electrons of the other spin orientation. (There will be a distribution, with the most likely spin being "down.")

Now suppose the left-hand detector is suddenly reversed. It now accepts only electrons whose axes point downward. Remarkably, the other detector will immediately pick up electrons whose axes spin upward. The transition will be instantaneous, as if information were transferred immediately from one electron to the other.

There seems to be a disturbing paradox here. How can one electron immediately know the orientation of the other electron's axis even if the electrons have been separated and no information can possibly be exchanged? That is the puzzle of Bohm's version of the EPR (Einstein-Podolsky-Rosen) paradox.

The standard resolution of the puzzle lies in the essential nonlocality of quantum mechanics. In the quantum picture of the world, one cannot truly speak of two separate particles. One must instead imagine that any particles involved in an interaction can be considered to be one wave function. This wave function serves as the fundamental quantity describing nature; one can no longer speak of position and velocity as the most basic natural measures. To put that in other terms: one cannot hope to construct a complete picture of reality at any given time. One can merely infer some of the properties of the wave function but not all of the positions and velocities of the particles themselves. Consequently, one cannot talk about the interaction between two particles as if the particles were completely separate. Rather, the particles are two "bumps" on the probability distribution picture of one wave function.

Needless to say, that resolution of the paradox has not been completely satisfactory to all. Some scientists feel the experiment is in fact a demonstration of supraluminal communication. They assert that information is being exchanged from one particle to another at a rate faster than the speed of light. This notion of "action at a distance" would clearly supersede Einstein's theory of special relativity if information could be sent at such a speed.

Some physicists are attempting to construct means by which messages could be sent at a velocity that is greater than the speed of light. In these schemes, one experimenter would monitor one detector and use it to send information to another experimenter at another detector. Most quantum theorists doubt that such instantaneous communication is possible. The problem is that in most schemes one needs to compare the results of the two detectors in order to see any correlation between the results. This verification procedure is a type of communication that is slower than the speed of light. Consequently, the entire process takes place at subluminal speeds.

The basic difference between the spin paradox experiment and standard sorts of communication lies in the notion of causality. Strictly speaking, the fact that one electron has an up spin does not *cause* the other electron to have a down spin; these are simply necessary coincidental events. It does indeed follow that the spins must be opposite, but it is not a matter of cause and effect. Both electrons come from a common source, a source which happens to contain one electron of each type of spin.

It is as if two shoes were drawn from the same shoebox. If you know that one of the shoes is for the left foot, then automatically the other shoe is for the right foot. That is not a case of cause and effect; it is simply an absence of other possibilities. (Unless, of course, the sales clerk has made a gross error, but that is beside the point!)

In our police car analogy, in which a row of cars is told to stop at a preset time, we find another example in which simultaneity might occur without communication. There would be no signal sent between the cars if that were to happen. Thus, communication could not be said to have taken place.

It would seem that the possibility of using variations of the spin paradox as communication devices is ruled out by such logic. Still, some workers in the field of quantum measurement claim that it is theoretically possible, under some circumstances, to send a message without any delay. A 1987 paper written by the Indian physicists Dipankar Home, Amitava Raychaudhuri, and Amitava Datta asserted that the speed-of-light barrier had been broken.[3] Although most scientists remain skeptical about the claim, there may be some grounds for hope: Perhaps quantum theory will someday make it possible to engage in instantaneous telephone conversations across the cosmos.

Synchronicity

It is hard to justify any claim that the synchronous behavior displayed by the spin paradox is the result of any causal connection between the two particles. On the other hand, some theorists have asserted that the behavior is a result of an acausal connection. The idea that acausal connections can exist is the hallmark of the system of beliefs, known as synchronicity, developed by the psychologist Carl Jung. Jung drew upon the works of quantum physicists to construct a theory of meaningful coincidences: so-called chance happenings which are related in a significant way. Jung saw synchronicity as a complement to causality applicable where determinism (causal connections between events) is not a valid approach:

> The philosophical principle that underlies our conception of natural law is causality. But if the connection between cause and effect turns out to be only statistically valid and only relatively true, then the causal principle is only of relative use for explaining natural processes and therefore presupposes the existence of one or more other factors which would be necessary for an explanation. This is as much to say that the connection of events may in certain circumstances be other than causal, and requires another principle of explanation.[4]

Jung saw incidents in his own life in which it was difficult to attribute certain connections between events to pure chance. For example, one day Jung experienced a series of events in which the image of fish played a part:

> I noted the following on April 1, 1949: Today is Friday. We have fish for lunch. Somebody happens to mention the custom of making an "April fish" of someone.... In the afternoon a former patient of mine, whom I had not seen for months, showed me some extremely impressive pictures of fish which she had painted in the meantime.. In the evening I was shown a piece of embroidery with fish-like sea-monsters in it.... I was at that time engaged on a study of the fish symbol in history.

With his interest triggered by examples in his own life and in the lives of his patients, Jung systematically studied what are called paranormal or parapsychological phenomena. He held that in many cases chance alone could not explain certain highly unlikely events.

The term "synchronicity" first appeared in a lecture given by Jung on the topic of the I Ching, the ancient Chinese book of changes. In this address, given in 1930, he related the concept of meaningful coincidence to the method of forecasting embodied in the I Ching. He asserted that causality is a western construct and that the Taoist philosophy of China is based on an alternative notion of the connection between events.

Although Jung mentioned the term "synchronicity" several times in his public presentations during the 1930s and 1940s, it was not until 1951 that he elaborated on what proved to be a highly controversial set of beliefs. In a brief lecture delivered that year to an audience in Ascona, Switzerland, he outlined his theory of "acausal connectedness." The following year, at the age of 77, Jung published his first and only manuscript on this concept, entitled "Synchronicity: An Acausal Connecting Principle."

The monograph was a result of a brief but fruitful collaboration between Jung and Wolfgang Pauli, one of the brilliant minds of quantum theory. As it turned out, each had something to offer the other. Pauli, who had psychological problems related to an inability to meaningfully engage in pastimes other than pure physics, came to Jung for psychoanalysis and dream interpretation. Upon completing his therapy, Pauli reciprocated by lending his efforts to Jung's program. Jung saw Pauli as someone who could provide a solid physical and mathematical foundation for the theory of synchronicity; he suspected that the quantum mechanical theory of action at a distance was somehow related to the notion of meaningful coincidence. Pauli, he thought, would help to make the parallels clearer.

Jung saw connections between the particle/wave duality of quantum mechanics and the duality between the unconscious and conscious minds. He was fascinated by the idea that thoughts and events could be related in unforeseen ways that were similar to the invisible connection between the electrons in the spin paradox experiment. Perhaps temporal and spatial distances do not serve as barriers for instantaneous communication within the subconscious mind?

It is interesting to compare this notion with the Australian aboriginal concept of dream time, a time and space in which the events of our dreams occur. One might also find analogies with Eliade's idea of sacred time, a time in which mythological events take place. In the Jungian perspective, dreams and myths represent links with shared universal archetypes. The connections seem to transcend ordinary space and time.

Jung's ideas have remained controversial and have won little acceptance in the field of psychology. Nor have they stimulated much of a dialogue between physics and psychology; the connections between quantum mechanics and the study of the mind, for which Pauli had hoped, have failed to materialize in a meaningful form. The parapsychological experiments on which Jung based his theory of synchronicity do not appear, to most workers in the field, to be statistically significant.

Even if ESP, telekinesis, clairvoyance, and the like, do not exist at all, it is of considerable interest to study the nature of mental communication and psychological time. The workings of the mind remain little understood. There are many well-documented cases in which mysterious forms of internal and external mental communication take place. It is hard for modern mental science to explain such occurrences.

One puzzle that continues to baffle most scientists is the case of the so-called human calculators or autistic savants (formerly called idiot savants). This remarkable group of people, although handicapped by an inability to communicate in ordinary ways, and in many cases also lacking in basic mental skills, possesses an extraordinary gift for calculation at a pace that can even, at times, rival that of computers. Many members of this group have difficulties with ordinary basic mathematics but can perform such feats as finding the square root of hundred-digit numbers. It is hard to characterize how these mental processes take place.

Consider the case of the twins, a pair of autistic savants well documented by the preeminent neurologist Oliver Sacks. Dr. Sacks records his astonishment at both the unbelievable mathematical abilities of these twins and the manner by which their astonishing calculations take place. For example, they could almost instantly determine the day of the week for any date in the last or next 40,000 years, seeming to work in harmony to complete this mysterious mental algorithm. Working in harmony, they could immediately count the number of matchsticks dropped on a floor. Yet their abilities appeared to be stunted whenever they were separated. Dr. Sacks recalls his second encounter with the twins:

> They seemed to be locked in a singular, purely numerical, converse. John would say a number—a six-figure number. Michael would catch the number, nod, smile and seem to savour it. Then he, in turn, would say another six-figure number, and now it was John who received it, and appreciated it richly....What were they doing?
>
> As soon as I got home I pulled out a table of powers, factors, logarithms, and primes....I already had a hunch, and now I confirmed it. All the numbers, the six-figure numbers, which the twins exchanged, were primes, i.e., numbers that could be evenly divided by no other number than itself or one....There is no simple method for primes of this order—and yet the twins were doing it![5]

The method employed by the twins to "compute" their astonishing results is truly unknown. It seems that there are mechanisms inside the human brain that can produce seemingly instantaneous calculations. Whether that represents a holistic way of thinking or simply an advanced algorithm beyond our present understanding is a matter of much dispute.

Even the rapid computations probably take place at a speed that is much lower than that of light. That is true of electronic computers also. Machines which rely on electric impulses to convey information are limited by the sublight speed of the electron, and for that reason there is an upper bound on their size. An electronic processor cannot be so large that there is a significant time lag between its components, which would slow calculations immensely.

Much work is being done with a generation of computers that would rely on *optical* communication: using light to relay messages. Fiber optics presents a wealth of new opportunities for rapid dissemination of information, for here the signals travel at the speed of light. With optical transmission of impulses, computers could be much larger and much faster.

To improve on optical communication, faster-than-light mechanisms would have to be invented. Naturally, the feasibility of such devices is highly controversial. There does, however, appear to be the possibility of large-scale synchronized motion. The motion, allowed by quantum theory, would take place instantaneously. Perhaps this approach can be exploited for the development of instantaneous communication on a small scale, namely, the scale of computers.

With an increased understanding, perhaps even the human mind will someday be fully understood. As Jung has pointed out, much that is going on in the mind appears to elude traditional science. It may indeed be true that synchronicity, rather than causality, is the operating principle behind human intelligence.[6] It is of considerable interest to explore the relationships between external time, the time measured by the clocks of the external world, and psychological time, the time of the hidden recesses of our minds.

References

1. Albert Einstein, *Relativity*, Crown Publishers, New York, 1961.
2. Richard Feynman, Robert Leighton, and Matthew Sands, *The Feynman Lectures on Physics*, Vol.III, Addison-Wesley, Reading, Mass., 1981.
3. Dipankar Home, Amitava Raychaudhuri, Amitava Datta, "A Curious Gedanken Example of the Einstein-Podolsky-Rosen Paradox Using CP Nonconservation," *Phys. Lett.* **A**, vol. 128 (1987), p. 4.
4. C. G. Jung, *Synchronicity*, Princeton University Press, Princeton, N. J., 1973.
5. Oliver Sacks, *The Man Who Mistook His Wife for a Hat*, Summit Books, New York, 1985.
6. F. David Peat, *Synchronicity—The Bridge between Matter and Mind*, Bantam Books, New York, 1987.

5
Shortcuts

The Flow of Time

When Einstein's theory of special relativity first appeared, it had an enormous, unprecedented effect on the general public. No other modern physical theory had captured the popular imagination in quite the same way. In one sense, the statement that clocks would run differently for different observers tapped into a general feeling of openness and "social and psychological" relativity. It was seen as confirming what many "knew" to be the case: that one cannot judge others by one's own standards.

The early part of this century saw a clear break with the social repression of the Victorian era, and the psychological theories of Sigmund Freud pointed the way toward a new appreciation of the complexity of the human mind. For many, Freud represented a sort of liberation from rigid social molds and outdated mores. Einstein's work also lent itself to this philosophy of social freedom, since many interpreted the theory of relativity to be a document which promoted a principle of social relativism that everybody should be allowed to pursue his or her own goals.

Moreover, the explicit program of Einstein—to show that time does not flow at the same rate for all—was seen as conforming to an intuitive sense that internal mental clocks function quite differently from the clocks on our walls. Time does seem to flow differently under different circumstances. For the young time seems to flow slower than for the old. It also appears to be true that, when one is bored, time crawls along at a snail's pace and, when one is pleasantly occupied, time whizzes past. Perhaps Einstein's theory can explain that.

Unfortunately, Einstein's theory cannot even begin to explain such phenomena. That points to one of the weaknesses of all modern physical theories of time: how can the psychological notion of time be entered into the picture? Einstein's theory of time is limited to a description of how clocks move under a variety of circumstances. It cannot be used to justify any notions regarding the human psyche.

The central mystery regarding the nature of time is that it appears to flow toward the future. It is hard to characterize this gut feeling that we all have that events are flowing past. We cannot say precisely where the world around us goes, because it crumbles and rearranges itself every instant. Are our temporal experiences analogous to watching a motion picture: the script has already been written, and the events have already been acted out; what remains is to watch the film of the proceedings? Or are the past and future simply incorporeal illusions? In other words, are we building the future and dismantling the past, or are we simply observing a predetermined set of events?

The standard response by physicists to such a line of inquiry is to point to the law of entropy as the moving force behind the psychological flow of time. We observe time's flow because of a perception that events can take place in only one direction. As we recall, Hawking has argued that psychological time must take place in the same direction as entropic time.

It is indeed obvious to all that certain chains of events cannot happen in a certain order. For instance, one would never see a pile of sand spontaneously arranging itself into a sand castle, but one might notice a sand castle being weathered into a pile of sand. The human mind naturally distinguishes between entropic and antientropic experiences. We expect disorder to increase, not decrease.

It is, however, hard to equate this perception of increasing disorder with the gut feeling that time flows onward; the feelings almost certainly have different origins. If one were to sit alone in a forest for several hours, one would still have a feeling that time is passing even if no visible irreversible changes (trees falling, etc.) were to occur during the period.

On the other hand, in the absence of decay or change, time would be meaningless. If one were condemned to spend eternity in a solitary white room without any possible form of diversion, it would make no sense to distinguish one moment from the next. In such a situation, time would have no arrow. Yet the true horror of such a situation would come from the fact that time would still be passing. The condemned person would have to experience an infinite number of indistinguishable instants one at a time. Thus, one must recognize that internal psychological time would continue to flow, even if all the conventional arrows of time were negated.

That is the thesis of the philosopher Adolf Gruenbaum: that one can distinguish between the arrowlike nature of time and the flowing behavior of time. Gruenbaum argues that the former is grounded in physical law, whereas the latter is not necessarily related to any external cause. The fact that time has a direction is independent of the fact that our existences appear to be plunging in that direction.

Gruenbaum, in *The Status of Temporal Becoming* clearly illustrates the difference. The arrowlike properties of time can be seen in a manner that is independent of what one means by "past," "present," or "future." For instance, if one considers a process that involves increasing entropy, as when a glass falls off a table and shatters, one can define an immistakable order of events. If one were to take snapshots of the occurrence, one could place them in their correct order without much problem. One could deduce that the events were anisotropic in time, in other words, that time has an arrow. The procedure could take place without reference to the ideas of "now," "earlier," or "later."

Gruenbaum suggests that the idea of "becoming" may be a mind-dependent construct. Time might not flow at all in the physical universe; the universe might simply be an entity consisting of unalterable events in space-time. Space and time would be equivalent in that scheme, but some mental process propels our attention from one time to another. The process represents a type of sensory perception similar to vision and smell. Thus, the flow of time would simply be an illusion and coming into being would be coming into awareness. As Gruenbaum states:

> Becoming is mind-dependent because it is not an attribute of physical awareness per se but requires the occurrence of states of conceptual awareness. These states of awareness register the occurrence of physical and mental events as sustaining certain apparent time relations to the states of awareness.[1]

The idea that the flow of time is unphysical in origin is echoed in the "block universe" idea of philosopher O. Costa de Beauregard, who argues that special relativity dictates a symmetry between the nature of space and time.[2] One can no longer speak of space and time as anything but concepts linked as the fusion called space-time. The fact that time flows is linked to human internal mechanisms rather than to objective physical reality. The moment of "now" has no significance whatsoever outside the realm of the mind.

Costa de Beauregard proposes that dreams and unconscious thoughts represent periods when the perception of time flow is suspended. During those times, one can experience the timeless reality of the block universe. It is only the conscious mind which is wedded to the illusion of temporal flow. The unconscious mind can run free and explore moments from the past with great ease.

As we can see, the views of Gruenbaum and Costa de Beauregard present a serious challenge to those who link the flow of time to the increasing entropy of the universe or to some other physical state. It is interesting to examine time's flow as a mental process, one that can be decoupled from the physical time as indicated by a clock. The flow can be hastened or slowed down by a variety of internal and external mechanisms.

Altered States of Temporal Experience

A child's approach to a typical day is quite different from that of an adult. For a child, a day seems endless, with hundreds of little games to play and tasks to perform. "Tomorrow" is something that a youngster rarely thinks about, and "next month" is forever. On the other hand, an adult might hardly notice the passing of a given single day. He or she might wonder about all the years flying by.

Certainly time appears to flow differently at different ages. What might seem like an excessively long interval for a child might pass exceedingly quickly for someone much older. Five years may seem like a lifetime for a child, whereas for a senior citizen it may be a drop in the bucket. One might wonder why that is so.

One answer to the riddle is that 5 years, which may seem a lifetime for a child, is, in fact, a lifetime for many small children! As we grow older, a year is a diminishing percentage of our lifetime. For someone who is 100 years old, a year is only 1 percent of his or her lifetime, whereas, for a 10-year-old girl, a year is 10 percent of hers. Consequently, time seems to pass one-tenth as quickly for the 10-year-old.

This answer is one of the standard approaches to the question of the hastening of time during the aging process. There is, however, strong evidence for an alternative explanation based upon chemical processes in the brain.

There is a certain amount of experimental support for the notion of a biological clock located in the brain. This chemically activated mechanism would regulate certain body functions. For example, the circadian rhythms of sleeping and waking are believed to be related to such a biological clock. Even when removed from exposure to the sun, humans and other animals continue to obey a fairly regular cycle of daily events. That is why many people can wake up at the same time each day and always anticipate an alarm clock by a few minutes.

A chemical pacemaker model of the human biological clock has been proposed by Hoagland and further developed by Treisman and Cohen.[3] In this model, a chemical pacemaker emits a regular series of signals that are recorded by a counting mechanism. Finally there is a third component which stores the number of pulses recorded by the counter. When inaccurate estimates of time intervals are made, there has been some error in the storage mechanism or some playback device. Presumably, external conditions such as temperature would affect the efficiency and rate of the clock.

One possible explanation of the effect of the aging process on time estimation lies in the possibility that the biological clock slows down upon the approach of old age. There is substantial evidence that many natural processes slow down considerably because of aging. That could conceivably affect the rate of an internal chemical pacemaker and create the illusion that time is speeding up because there would be fewer "ticks" per minute.

The biologist Du Nouy set out to find a means by which the speed of the internal physiological clock could be ascertained. He chose, as a measure of the rate, the time of cicatrization of surface wounds; in other words, he looked at how long it would take for wounds of a certain size to heal. He performed the same experiment for patients of 20, 30, 40, and 50 years of age. Averaging over the groups, while rejecting the results for clearly unhealthy patients, he obtained age-dependent values for the rate.

Du Nouy's results are rather remarkable. The time of healing increases substantially because of the aging process. For instance, if a child of 10 would take 20 days for a wound of a certain size to heal, a man of 30 would take over 40 days, and a man of 60 would take 100 days. The rate increases uniformly and roughly exponentially.

Because of the slowing down of bodily processes such as cicatrization, the time needed to perform a certain amount of bodily repair grows much longer as one gets older. In other words, for the same amount of work to occur, a greater amount of actual time needs to pass. Time, therefore, seems to pass much more quickly as the years go by. Du Nouy has concluded that: "Everything occurs as if sidereal (stellar) time flowed four times faster for a man of fifty than for a child of ten.... Thus, we find that when we take physical time as a unit of comparison, physical time no longer flows uniformly."[4]

Du Nouy thereby argues for a sort of "biological special relativity." External time, he proposes, is without beginning or end, and cannot be said to flow at all. Physiological time, on the other hand, seems to flow with different velocities for different observers. There is no objective biological time; there is only relative time based upon the aging process. Since we cannot come into contact with true physical time, physiological time is the only flow of time of which we can speak.

Undoubtedly, the situation with regard to the causes of and relationship between physiological and psychological time is far more complex than the chemical pacemaker scenario. Du Nouy's work has shed some light on the relationship between the flow of time and aging, but there are many other factors which can affect time's flow. It is of great interest to understand the interplay among all these influences.

A sizable amount of work has been done with regard to the experimental psychology of time. Most of the studies are related to the estimation of short durations. Typically, the estimates are made by presenting the experimental subjects with a certain fixed time interval which could, for example, be the time between two musical notes. Subjects must then either judge the length of the interval numerically or reproduce the interval themselves. The subjects' estimations of the time span are then compared with the actual value. The experiments take place under a variety of circumstances, and the conditions are closely monitored in order to measure the influence of a single parameter on time duration estimation.

One of the most interesting experiments that has been performed is concerned with a comparison of the time estimation of so-called "empty" intervals and intervals filled with various diversions. In a typical setup, subjects are asked to record the amount of time for a period in which there are no distractions. Then they are requested to sit for a second period of time in which there is much going on in the form of visual or auditory stimulation. What is found, typically, is that "filled" intervals seem to pass quicker but are remembered as being longer than "empty" intervals. In other words, if one can recall a number of events associated with a given period of time, that period, in retrospect, seems longer than it actually was.

The effect measured in these experiments is related to the common experience that, during boring or inactive periods of one's life, time seems to pass painfully slowly. During these periods, one might even try to kill time — to make the hours go by faster by the use of diversions. On the other hand, eventful periods of one's life seem to pass by much faster. During these times, each moment seems unique and precious. Consequently, the stretches are more easily remembered than the boring periods. Thus one is faced with the irony that a busy hour flies by quickly but may appear long in hindsight, whereas a boring hour goes by slowly but may appear insubstantial in retrospect.

Another interesting set of psychological observations relates time interval measurement to the estimation of distance and speed. These investigations have led to the discovery of the so-called tau and kappa effects. The tau effect relates to the determination of spatial intervals when time intervals are given. It was first discovered in a series of tactile experiments. Subjects were asked to estimate the distances between a set of three points marked off on their skins. The second point was equidistant from the first and third points, but the time taken by the second and third stimulations was longer than that by the first and second stimulations. Consequently, subjects mistakenly supposed that the distance between the second and third points was greater than that between the first and second points.

The conclusion from that series of experiments was that, if the time between sensory stimulations is greater, the distance between the stimulations is often overestimated. Similar results have been found for visual and auditory signals.[5]

The converse of the tau effect is called the kappa effect. Here the distances between sensations are varied while the time between them is kept constant. The subject typically supposes that, if the distance between two sensory stimulations is greater, the time between these events is greater — even if that is not so.

The kappa and tau effects are well documented. They reveal the full extent to which, in psychology as well as in physics, the concepts of time and space are closely related. This relationship is especially of interest when one studies the way children appreciate certain physical concepts such as speed and distance. The psychologist Jean Piaget has examined how those attitudes change as children grow. Before the age of 8, most children believe that if an object travels a great distance, the time taken must be correspondingly greater. It is only after the age of 8 that children intuitively comprehend velocity.[6]

The estimation of both temporal duration and spatial distance may be strongly affected by the use of psychoactive drugs; certain chemicals may profoundly alter the rate and manner by which information is perceived by the senses.[7] The effect of the psychoactive substance LSD differs greatly from person to person. For that reason, it is hard to generalize about the temporal and spatial distortions produced by the drug. Nevertheless, some common features are associated with the hallucinatory experiences induced by the intake of LSD. First, space seems to lose its regular, Euclidean character. Certain images may loom larger than others, depending upon their significance to the experimental subject. Generally, space appears to stretch out. Time, on the other hand, seems to be accelerated. Moments pass by at a much faster rate. It is unclear how LSD affects the biological clock and why its effects are so varied.

Tranquilizers produce the opposite effect. Under the influence of one of those substances, time appears to stretch out. What would seem like a short time to a person in an ordinary frame of mind seems much longer to someone who is tranquilized. No pharmaceutical chemicals known to humankind slow down the body's internal clock by a significant amount. Consequently, there are no drugs which are taken specifically to alter the physiological time of the user. It is interesting, though, to consider the effects that such a drug would have if it were to be developed.

In "The New Accelerator," a short story by H. G. Wells, the idea of using drugs to speed up or slow down human metabolism is fully explored. In the story, a professor of chemistry develops a chemical which speeds up the biological clock of anyone who ingests it. On testing the drug, the professor notices that his own physiological time speeds up by a factor of 1000. His motions become extraordinarily rapid, enabling him to perform amazing feats of speed and dexterity. While under the influence of the accelerator, all the world around him seems to slow down. Insects hover in front of him, suspended in midair, while horses seem frozen in their tracks. People stand like statues all around him. Eventually the drug wears off and everything appears normal once more.

The professor goes on to develop another drug, called a retarder, which slows down the human physiological clock by a substantial amount. Users of this substance perceive that the world around them is moving far more rapidly than it actually is. The protagonist of the story foresees a great market potential for these chemicals:

> The two things (accelerator and retarder) together must necessarily work an entire revolution in civilised existence. It is the beginning of our escape from the Time Garment of which Carlyle speaks. While this Accelerator will enable us to concentrate ourselves with tremendous impact upon any moment or occasion that demands our utmost sense and vigour, the Retarder will enable us to pass in passive tranquillity through infinite hardship and tedium.[8]

The story presents us with the interesting possibility that physiological time may be speeded up or slowed down at will. The chances that accelerators and retarders will be developed are quite remote, however. The human body could not operate at a much higher metabolism without serious health consequences. Also, it is unclear what the full impact of any change in the rate of physiological processes would be upon the psychological perception of time. Any acceleration in the body's metabolism would most likely lead to exhaustion in short time without any significant impact upon time perception. Nevertheless, it is fascinating to imagine what the results of such a bold experiment would be, if it were in fact realizable.

The simplest way to alter the flow of time is a quite common one: by sleeping. Sleeping represents a clear break in the normal flow of time. Some scholars, such as Jung and Costa de Beauregard, have even likened sleep to a state of timelessness. Many people consciously use sleep as a means of escaping from time's flow. It provides a reliable means of pushing the external clock forward without having to experience the intermediate moments. In addition, it leads to a sense of renewal and, in some small way, rebirth.

Sleep does not, however, represent a pure state of timelessness. It is often interrupted by dreams. During dreams, time does seem to flow. However, unlike time's steady rhythm during the waking hours, dream-time passes in a strange and sometimes surrealistic manner. According to the traditional Freudian interpretation, dreams originate from the repressed wishes buried in the unconscious mind. The unconscious mind is, by nature, timeless. Therefore, dreams do not inherently possess a flowing form of time. When time does appear in a dream, it is of significance primarily as a reflection of the sleeper's current responsibilities in the waking world. The rate of time's flow during part of a dream reflects the sleeper's attitude toward the symbolically depicted events that the dream represents. For example, if time seems uncomfortably fast-paced during a dream about a sleeper's work life, some anxiety is probably associated with the work.

Most of us can recall only a small fraction of our dreams. When we do remember our dreams upon awakening, we tend to rearrange and spread out the order of events in order to make sense of our experiences. If a lot of events are associated with one part of a dream, we might assume that the time taken for them is longer than for an uneventful segment. Thus, a few seconds of dream time can seem like hours of real time.

Detailed experiments to measure the correlation between the amount of time a dream seems to take and the actual clock time of the dream have been performed. The fact that there is a significant link between dreaming and rapid eye movement (REM) aids in the understanding of how long a dream takes to occur. It has been found that a subject usually is dreaming during periods in which his or her eyes turn quickly from side to side. The correlation provides us with an estimate of the length of the periods. The studies show that there can be a significant discrepancy between the imagined and actual times of a dream.

The time distortion embodied in dreams can be reproduced through clinical hypnosis. A hypnotized subject can be influenced to perceive time as moving faster or slower than a clock indicates. For instance, a suggestible person (one who can easily be hypnotized) can be told that half an hour will pass between two events. If all external time devices are removed, the subject may believe that 30 minutes have passed when the actual time is longer or shorter. That leads one to conclude that the experience of the flow of time has strong psychological components which can readily be influenced by suggestion.

Many other factors can influence the rate at which time seems to pass. Emotions can strongly influence the rate of temporal passage; for example, grief can make an hour seem unbearably long. Cultural influences also can play a part in determining the rate of one's inner clock. In highly developed industrial societies, psychological clocks must operate at faster rates than in slow-paced agricultural societies. Finally, climate can play a role in determining the speed of internal time. In colder areas, biological clocks tend to be slower than in more temperate areas.

There are also many organic causes of changes in the flow of time. For example, one might consider the effect of the processes that take place before death or near-death incidents. A commonly held belief is that, before death, one's life "flashes before one's eyes." There are many documented cases in which this phenomenon has occurred during a near-death experience. Scientists are unsure of the origin of the break in the normal flow of time. It has been postulated that it may be caused by the deterioration of brain tissue.

Many psychotic patients are said to experience radical alterations in their perception of temporal processes. For example, there is a strong tendency among schizophrenics to underestimate their age or period of confinement. Some schizophrenics even live in a world without time. As Meerloo points out:

> Perhaps [there is] a typical schizophrenic time experience [consisting] of living in a timeless archaic world without rhythm, without night and day, in a kind of oceanic time such as the fetus experiences in the womb. Many schizophrenics think about their psychotic phase as an eternity. They are aware of the loss of time consciousness.[9]

Disease can play a strong role in distorting time. Certain types of brain disorders can lead to a drastic loss of memory. Clearly, an abrupt memory loss distorts the perception of time by a considerable amount. A victim of such a syndrome might lose the ability to distinguish between events of the distant past and those of the present. Time might even cease to flow for such a person.

In 1887, the Russian physician Korsakov documented a strange mental disorder that he believed to be related to alcohol-induced brain damage. The disorder, now called Korsakov's syndrome, is typified by severe memory loss due to neuron destruction in the brain. Sufferers generally maintain a significant fraction of their long-term memory but lose much of their short-term memory. Thus, there is a general sense of being trapped in time. Quite often, patients are not aware that anything is seriously wrong with them except the vague impression that something is not quite right.

Oliver Sacks documents several cases of Korsakov's syndrome in his book *The Man Who Mistook His Wife for a Hat.* He relates that "such patients, fossilized in the past, can only be at home, oriented, in the past. Time, for them, has come to a stop." One of his patients, called Jimmie, has experienced a severe trauma due to alcohol. Diagnosed with Korsakov's syndrome, he seems locked in the past. He remembers nothing after 1945 and almost instantly forgets anything told him. For him, World War II has just ended even though the calendar reads that it is 4 decades later. Seeing his brother, who of course has aged, he cannot believe this "instant" transformation in a man that he believes is still a teenager. Jimmie must continuously discover where he is and what he is doing, for he cannot maintain any kind of continuity.

Another of Sacks's patients, a man named William Thompson, developed acute Korsakov's syndrome after a brief period of high fever. He lost the ability to hold a thought in his mind for more than a few seconds; consequently, he had to make up his world every moment of his life. He is constantly engaged in story telling to make sense out of his shattered life. Moreover, although he exists in a constant state of anxiety, he is unaware of any loss. Sacks recalls one of his encounters with Mr. Thompson:

> He remembered nothing for more than a few seconds. He was completely disoriented. Abysses of amnesia continually opened beneath him, but he would bridge them, nimbly, by fluent confabulations and fictions of all kinds. For him they were not fictions, but how he suddenly saw, or interpreted, the world....So far as he was concerned, there was nothing the matter...[He] must literally make himself and his world up every moment.[10]

The cases described by Sacks raise serious questions about the nature of time. In particular, they show us how delicate is the connection between psychological time and physical reality. The linkage depends upon inherent and learned associations, between external events and our thoughts, built up over the years in our brains. Once the connections are severed, it is hard to talk about time as having a flow. Each moment becomes an independent snapshot of a timeless block universe.

Would the flow of time still exist if all of us lacked the facilities to weave our momentary perceptions into our tapestries of memory, or could the universe be said to be timeless in the absence of record-keeping, conscious observers? Is the experience of time's flow a product of our mind's ability to make sense out of disparate events, or is it a true representation of the deep nature of reality? The question of whether or not time is a human construct cannot be directly answered. It remains a profound mystery, one that will probably elude humankind forever.

Time Travel

Episodes in which the fabric of time is altered or the flow of time is disrupted provide us with an interesting alternative approach to temporal awareness. One can imagine time to be something that is fluid and subjective rather than rigid and universal. In this dynamic version of time, one can even imagine separating the time line of a person from the time line of the world. In other words, it may be possible for someone to escape from the confines of clock time altogether. That would present the possibility of traveling through time.

Time travel has been a common theme in much speculative literature. The notion that one can withdraw from time's flow and rejoin at a later or earlier moment is a most fascinating one. Time travel provides a reprieve from one of the unfortunate conditions of mortality: we can neither experience the far future nor relive the past. Although with time travel one cannot escape death, one can significantly extend the fraction of human history that one has the opportunity to glimpse. In a traditional time travel story, the protagonist is transported into the past or future (usually the future) by some device. For early time travel stories, the common means of transport are some of the methods of altering the flow of time that have been mentioned in the previous section, namely, sleep, dreams, drugs, hypnosis, and memory lapses.

In many of the early stories involving travel to the future, sleep is the favorite technique for transport. In *Memoirs of the Year Two Thousand Five Hundred,* written by L. S. Mercier in 1771, a man falls asleep and wakes up in a utopian community of the distant future. In another example of this kind of story, Rip Van Winkle, the protagonist of Washington Irving's story of the same name, sleeps for a much shorter period of time and awakens to see his aged relatives.

It is harder for an author to explain travel to the past by use of sleep. Nevertheless, in Mark Twain's *A Connecticut Yankee in King Arthur's Court,* a blow on the head propels the hero into a sleeplike state. He awakens in the past, having traveled from America to medieval England.

When hypnosis, or mesmerism became popular in the nineteenth century, it supplemented sleep as a supposed means for time displacement. In *Looking Backward,* Bellamy's future Bostonian utopia is experienced by a man who has been placed in a deep trance. While the protagonist is sleeping soundly in a subterranean crypt, his house burns down. Falsely presumed dead, he is excavated by later occupants of the site.

A substantial shift took place in the format of time travel stories with the publication of *The Time Machine* in 1895. Wells's tale depicted how someone could travel into the past or future at will rather than simply through uncontrollable external forces. Wells imagined a mechanism which would enable someone to travel along the fourth dimension, time, as easily as one could travel along the three spatial dimensions. As the reader will recall, Wells's notion of space-time predated Einstein's formulation of the concept.

By adjusting the dials of his machine, Wells's time traveler can journey to any era of history; his device can travel, with equal ease, to the past or to the future. The sensation that he experiences while traveling is that of a continuous thrust forward. All around him, people seem to move by at incredible speeds. Buildings appear and disappear all around him. The sun and the moon pass like bands of light across the sky. The experience is truly extraordinary.

The time traveler's biggest fear is that he will land in a location that is already occupied. Were he to land in such a spot, his molecules would crash into the molecules of the other object and there would be an enormous explosion. The time traveler is terrified by that possibility:

> The peculiar risk lay in the possibility of my finding some substance in the space which I, or the machine, occupied. So long as I travelled at a high velocity through time, this scarcely mattered: I was, so to speak, attenuated and was slipping like a vapour through the interstices of intervening substances! But to come to a stop involved the jamming of myself, molecule by molecule, into whatever lay in my way: meant bringing my atoms into such intimate contact with those of the obstacle that a profound chemical reaction, possibly a far reaching explosion, would result.[11]

Fortunately for the time traveler, no such accident occurs. It is interesting that there is not even an effect due to the inevitable collision with air molecules that must result upon halting the machine. The time traveler does not even consider that possibility.

Wells's protagonist does not attempt to travel into the past. If he were to do so, there might be the possibility that he would create a time paradox, which occurs when a traveler through time alters the past in a significant manner. Upon returning to the future, he finds that the world in which he grew up has been changed, possibly even beyond recognition.

There are numerous time travel stories involving time paradoxes. One of the most famous of them is "The Sound of Thunder," by Ray Bradbury. In this story, a "time safari" company sends adventurous hunters into the past to hunt for dinosaurs. The hunters are carefully told where they may tread and which animals they may kill. Those warnings are necessary to prevent any alteration in the events of the past that may have future ramifications.

On one such expedition, the warning is unheeded. A butterfly is accidently killed. The effects of that death magnify through time (a primitive example of the "Butterfly Effect" of Edwin Lorenz mentioned in Chap. 3!) and lead to disastrous consequences. A pivotal U.S. presidential election's outcome reversed, and that leads to unpleasant results worldwide. All that happens because of a minor disruption in the delicate chain of cause and effect that has molded the world of the story's present time.

In the above story there really isn't a paradox, in the strictest sense of the word, because there isn't a real contradiction between two opposing principles. In other time paradox stories, however, the reader is confronted with a clash between several contradictory versions of reality that cannot be reconciled with one another. The classic example of such a paradox is a turn of events that leads to a woman murdering one of her own ancestors. The loss of an ancestor would mean that the woman never existed and that, consequently, she could never have traveled back in time. Therefore, her ancestor is not killed, so she really is alive and can perform the murder, and so on. There is no way to resolve the paradox: One cannot assume either that the time traveler is alive or that she is dead. In either case, there is a contradiction.

In numerous variations on this theme, time travelers disrupt causation in a variety of ways. One interesting subclass of such stories involves events which cause themselves. In "All you Zombies," by Robert Heinlein, a man goes into his past and by a twist of fate becomes his own father and (through a sex change operation) his own mother! Therefore, something seems to be created from nothing, since the man is his own parent.

Another interesting possibility is that a time traveler sends plans for a time machine into the past. Someone in the past uses the plans to perfect time travel. The time machine is invented and is used to propel the plans for the machine's design back into the past. Again something (the time machine) is created from nothing.

There are several methods by which authors propose to resolve time travel paradoxes.[12] One way to avert paradox is to postulate that the universe has a certain resilience. Any alterations in the web of cause and effect are smoothed over by subsequent changes. Another is to propose the existence of alternative universes. Whenever a change is made by a time traveler, the time traveler is sent into an alternative universe in which there is no contradiction. Suppose for example, someone kills his own mother. If he tries to go back to the present, he finds that he is in another version of the universe, one in which he never existed. We shall deal more with the question of alternative universes in the next chapter.

So far, our discussion has been completely speculative; we have been analyzing something which could be a complete fiction. At this point, we might wonder if time travel of the Wellsian sort could ever be realized. Is there any known way of traveling through time that conforms to physical law? Or is time travel simply a fantasy?

The answer to these questions is rather complex. We must consider separately the possibilities of travel into the future and travel into the past. Travel into the future may be far easier than returning to a past era for the simple reason that we are already moving into the future. Our existences travel into the future at the rate of 24 hours a day. By proposing travel into the far future, we would want to speed up the flow of time rather than reverse it. Reversing time, in order to travel into the past, certainly is more difficult and may be impossible. Let us examine why that is so.

The simplest way to travel into the far future is to exploit the theory of special relativity. Whenever an object travels at a certain velocity relative to the earth, its internal clock appears by earthbound standards to slow down. Consequently, the object can be said to age less than it would if it were attached to the earth.

For example, suppose a spaceship were to travel away from earth at a velocity that is close to the speed of light. A passenger on the ship, Ted, begins the journey at 28 years of age. His brother, Ned, who remains on earth, is also 28 years old. Suppose that there is a television monitor by which Ned can watch the goings-on of Ted.

As the spaceship reaches its maximum velocity, Ned notices that Ted's actions have become slower and slower. Eventually, Ted seems to be in a state of suspended animation. Ned ages and eventually dies. Ted returns to earth after what has been, for him, a 7-year period. He finds that several hundred years of earth history have passed in the meantime. The result is that he has traveled into the future of earth without having aged significantly.

This motif has been used in a number of time travel stories. Typically, a team of astronauts leave earth, travel close to the speed of light, and then return several centuries later. This is the means of time travel that is used in adventures such as *Planet of the Apes,* in which an astronaut returns to a future earth that has been horribly altered.

The use of special relativity as a means of time travel is a very real possibility. Relativistic time dilation is a phenomenon that has been scientifically verified. It has been measured, for instance, on clocks that have been sent up in fast airplanes. Of course, the speed of even the fastest airplane is still much less than that of light, so all relativistic effects are extremely small. No one, however, has as yet traveled at speeds comparable to that of light. Therefore, there have been no significant human experiments involving the time dilation effect.

It is unclear what would happen to the human subjective clock as the body is brought to near-light speeds. It is assumed that subjective time would mimic the clock time and accordingly, slow down relative to the subjective time of somebody who is on earth. However, no one knows for sure what would happen to our internal clocks.

In the short story, "Common Time," by James Blish, the scenario is one in which subjective time lags behind relativistic clock time. A space traveler discovers, to his horror, that his psychological time is running thousands of times faster than the clock time would indicate. Thus, every second of clock time corresponds to 2 hours of thoughts. Fortunately, the effect wears off as the traveler contemplates years of complete boredom. Hopefully, such an effect does not occur as near-light speeds are reached. If subjective time does correspond to relativistic clock time, it would indeed be possible for the mind, as well as the body, to travel through the ages.

Notice that the time travel allowed by special relativity must take place into the future. Relativity can alter the flow of time but cannot change the order of events. It cannot switch the sequence of cause.and effect. Thus, past time travel is impossible by this method. In order to journey into the past, one would have to travel *faster* than the speed of light. That is an eventuality excluded from the theory.

This exclusion hasn't stopped writers of speculative fiction from utilizing faster-than-light travel as a means of returning to the past. By breaking the "light barrier," the protagonists of such stories find that their clocks begin to run backward. Eventually they can reach any time in the distant past by use of the method. This pseudopossibility has also inspired some humorous poetry. The following famous poem about faster-than-light travel has appeared anonymously:

> There was a young girl named Bright
> Who could travel much faster than light
> She went out one day
> In an Einsteinian way
> And returned the previous night.

Although special relativity specifically excludes time travel to the past, general relativity does not. In general relativistic theory it is technically possible for a strong gravitational field to alter the space-time fabric. If the fabric is distorted significantly, time travel into the future or past might be possible.

Let us review what we have discussed about general relativity and black holes. According to general relativity, massless space-time can be represented by a four-dimensional surface. The presence of mass distorts the surface much as a trampoline is distorted when someone stands on it. The more mass that is present, the more space-time that is warped.

The warping of space-time affects the motion of all objects in the vicinity. Normally, we associate the change in motion with the effect of gravity, but here we find the additional result that *massless* objects' motions also are affected. In traditional Newtonian gravitational theory, massless objects should not experience gravity. In Einsteinian general relativity, massless objects, such as light particles, *are* influenced by gravity.

As we have pointed out, black holes are extremely dense byproducts of stellar death. Their masses act to distort the space-times around them considerably. One result is that they are surrounded by so-called event horizons. Once inside an event horizon, no object can escape intact. Even light must be trapped by the "semipermeable membrane." Current theory holds that energy does leak out of a black hole but is a different form than the matter and energy that enter it. Consequently, one can assert that matter does not leave the black hole in a recognizable form.

In the center of the black hole is a point in space-time, called a singularity, that represents a break in space-time. Once passing into a singularity, an object passes over the "edge" of space-time. By the time it reaches the singularity, however, it has been squeezed and stretched by the black hole's enormous gravity. As a result, it has become infinitesimally thin before entering the singularity.

If a black hole rotates, something interesting happens. According to theory, there is a region inside a rotating black hole where time and space exchange roles. That is because of the tremendous distortion of space-time caused by the black hole. Consequently, in this region one can travel through time freely; one can reach any point in the past or future. Conceivably, such black holes could be used for travel through time.

The state of affairs inside this "timeless" region of a black hole is the exact opposite of that of normal space. In the world outside a black hole, one can travel freely through space but not in time. Inside the black hole, spatial movement is restricted but temporal motion becomes possible.

One might wonder how feasible this method of time travel would be. Theoretically, this sort of time displacement can take place in a relatively straightforward way. In practice, any astronaut caught in a black hole would run the very serious risk of being crushed. Only in extremely large black holes could the risk be somewhat minimized.

Second, there is only one theoretically postulated method of escape from a black hole. It utilizes the highly speculative idea that black holes provide gateways to other universes or other parts of our universe. This controversial notion was spawned by mathematical models of black holes in which there seem to be connections between two universes via a "wormhole."

So someone seeking to travel through time could enter an extremely large (galactic size) rotating black hole. He then could possibly reach a region in which temporal displacement is possible. Carefully steering away from the crushing gravity of the center, he might be able to escape through a wormhole. Conceivably, he could then travel back to earth, arriving sometime in earth's past or future. This series of events, though theoretically possible, is just not very likely or practical. Therefore, we might rule out travel into the past as something that is grounded in physical theory.

In summary, time travel seems to be possible when the aim is to voyage into the future. Special relativity allows for this interesting eventuality, which someday might be realized. Travel into the past is much more problematic and may be, in fact, impossible.

In addition to the physical method of time travel, other means of temporarily escaping time's flow are quite accessible. The simplest of the methods is sleep. Long-term sleep, through suspended animation, may someday be practical. Other methods include drugs and hypnosis. Certain types of mental and neurological disorders sometimes cause time travel of a different, and usually unwanted, sort, as in the case of schizophrenia.

It is clear that time is not a uniformly flowing fluid but can have many eddies and ripples. The flow of time depends largely, if not entirely, on the observer. In the next chapter, we shall explore the possibility that time has, in addition, many branches.

References

1. Adolf Gruenbaum, *Modern Science and Zeno's Paradoxes*, Wesleyan University Press, Middletown, Conn., 1967.

2. O. Costa de Beauregard, "Time in Relativity Theory—Arguments for a Philosophy of Being," in J. T. Fraser (ed.), *The Voices of Time*, George Braziller, New York, 1966.

3. John Cohen, "Time in Psychology," in Jiri Zeman (ed.), *Time in Science and Philosophy,* Elsevier Publishers, New York, 1971.

4. P. Lecompte Du Nouy, *Biological Time,* Macmillan Publishing Co., New York, 1937.

5. John Cohen, "Subjective Time," in J. T. Fraser (ed.), *The Voices of Time,* George Braziller, New York, 1966.

6. Jean Piaget, "Time Perception in Children," in J. T. Fraser (ed.), *The Voices of Time,* George Braziller, New York, 1966.

7. Joost A. M. Meerloo, "The Time Sense in Psychiatry," in J. T. Fraser (ed.), *The Voices of Time,* George Braziller, New York, 1966.

8. H. G. Wells, "The New Accelerator," from *28 Science Fiction Stories of H. G. Wells,* Dover, New York, 1952.

9. Joost A. M. Meerloo, "The Time Sense in Psychiatry," in J. T. Fraser (ed.), *The Voices of Time,* George Braziller, New York, 1966.

10. Oliver Sacks, *The Man Who Mistook His Wife for a Hat,* Summit Books, New York, 1985.

11. H. G. Wells, "The Time Machine," in *Three Prophetic Novels* selected by E. F. Bleiber, Dover, New York, 1960.

12. Martin Gardner, *Time Travel and Other Mathematical Bewilderments,* W. H. Freeman and Co., New York, 1988.

6
The Garden of Forking Paths

Time as a Labyrinth

During the nineteenth century it was somewhat common in Britain for owners of large estates to construct mazes in their gardens. By carefully trimming the hedges in a formal garden, intricate and baffling labyrinths were designed. Visitors could amuse themselves by trying to solve the mystery of a maze. They could lose themselves in the thicket, then try their luck at leaving the maze by the shortest way possible.

The writer Jorge Luis Borges had a curious passion for labyrinths, and he also had a keen interest in the puzzle of time. For Borges, it was not clear where the dividing line between these two preoccupations was. All his life, he wished to understand the riddle of time as if he were attempting to find his way out of an infinite maze. Borges considered time to be the greatest mystery, the puzzle of puzzles.

Perhaps no other story depicts the labyrinthine character of time in clearer detail than Borges's "The Garden of Forking Paths." In this story, time is seen as the answer to a riddle of infinite complexity. It is a maze from which there is no exit.

The story takes place during World War I. Yu Tsun, a Chinese spy who is working for Germany, is attempting to elude capture by the British. He finds refuge in the house of a noted Sinologist, Stephen Albert. Albert impresses Yu Tsun with his knowledge of one of Tsun's illustrious ancestors. This ancestor, named Ts'ui Pen, a provincial governor, astronomer, interpreter, calligrapher, and poet, was quite an influential man. One day he announced suddenly that he was giving up his power. Abandoning his wealth and status, he took on the life of a recluse.

125

When asked why he made that drastic move, he replied that he was withdrawing to compose a book and a maze. He continued in his monastic existence for 13 years, until his death. When he died, fragments of a chaotic manuscript were found. The jumbled manuscript was published in accordance with Ts'ui Pen's will.

Scholars could not unravel the mystery of the manuscript, nor could they locate the labyrinth that Ts'ui Pen had created. Finally, Stephen Albert came across a letter that Ts'ui Pen had written shortly before his death. In this letter there was a statement that revealed Ts'ui Pen's intent: "I leave to the various futures (not to all) my garden of forking paths."[1] Upon reading this letter, Albert solved the mystery of the book and the labyrinth: they were one and the same. Ts'ui Pen's manuscript could be identified as his "garden of forking paths."

Ts'ui Pen's novel is quite unusual. Characters are killed and later reappear. Elements of the plot are blatantly contradicted as the novel progresses. In effect, the book contains all eventualities; instead of the characters in the book choosing one of many alternatives, they choose every possibility. Each of the various futures is depicted. Thus, the novel is a labyrinth in which a reader becomes lost in the variety of options.

After much contemplation, the purpose of the book has become apparent to Albert. Ts'ui Pen was obsessed with the meaning and form of time. He wished to convey the complexity of time's mystery in the form of a riddle. In a riddle, the only word that must not appear is the answer. Similarly, in Ts'ui Pen's book, the word "time" did not appear at all. In Albert's extensive studies of Ts'ui Pen, he has concluded that the nature of time must have been the focus of Ts'ui Pen's writings. He asserts to Yu Tsun that Ts'ui Pen had a radically unconventional approach to time:

> "The Garden of Forking Paths" is an incomplete, but not false, image of the universe as Ts'ui Pen conceived it. In contrast to Newton and Schopenhauer, your ancestor did not believe in a uniform, absolute time. He believed in an infinite series of times, in a growing, dizzying net of divergent, convergent and parallel times. This network of times which approached one another, forked, broke off, or were unaware of one another for centuries, embraces all possibilities of time. We do not exist in the majority of these times; in some you exist, and not I; in others I, not you; in others, both of us.[2]

Yu Tsun thanks Albert for his insights and then kills him for military reasons, thus confirming the many-forked aspect of time. Like the characters in Ts'ui Pen's book, two contradictory notions, that Tsun and Albert are both friends and enemies, are realized at the same time.

Borges's story provides us with insight into another model of time. In this picture, time is seen as constantly forking. The universe continually branches into various subdivisions whenever a choice is made. The branches can separate or come together at various junctions. This image, which can be called the many-forked model of time, can be used to remedy some of the problems associated with time travel paradoxes. Consider, for instance, the case in which a person goes back in time (before her mother's birth) and kills her maternal grandmother. If time is a single strand, there is an obvious paradox: If someone didn't have a mother, how could she exist in the first place? A multiforked model of time removes this contradiction. The time-traveler's mother would exist in one of time's branches and not in another. By preventing the birth of her mother, the time-traveler transports herself from one branch of time to another. Hence there is no paradox.

The idea of alternative universes that lie parallel to our own has appeared in many science fiction stories. The writer who has been most closely associated with parallel universe stories is the late American author Philip K. Dick. In his novel *Valis* he suggests that modern history is a distorted version of the true development of the world. Somehow, "evil forces" have shunted the world along an unlikely fork in time. In *The Man in the High Castle,* an alternative picture of World War II is presented. Germany and Japan have won the war and have divided up the world. It is made clear in this story that the scenario depicted is that of a parallel branch of our universe, characters may even travel "sideways in time" from one branch of the universe to another.

The speculative nature of time travel might lead one to believe that the idea of alternative universes borders on science fiction. It would seem that there is no physical basis for these speculations. There is no way of proving or disproving the existence of alternative universes if it is impossible to gain access to them.

The Mystery of Schrödinger's Cat

Quite recently there has been a serious attempt by some physicists to incorporate the many-forked model of time into physical theory. Motivating this attempt is a deep-seated feeling among many theorists that the Copenhagen interpretation of quantum mechanics is not a wholly satisfactory explanation of the world of particles. We know that many physicists have had trouble accepting some of the fundamental notions of quantum mechanics and that one of these troublesome concepts is the idea of wave functions. Let us examine how the concept of wave functions has supplanted many traditional, intuitive notions about the universe.

In the classical Newtonian picture of nature, all particles in the universe were said to have definite properties, such as position and velocity, which possess unequivocally measurable values. The values can be confirmed by repeated observation. In any observation, one can break the world up into two parts: observer and observed. It is manifestly clear where one part ends and the other begins. In any scientific experiment, it is assumed that the interference of the observer is minimal.

In the current, quantum model of particle interactions, such observables as position and velocity are not seen as fundamental. Instead, the wave function, which is an abstract entity that cannot be measured directly, is the basis of all knowledge of particle behavior. The wave function need not be in a state of definite position or velocity. In general, the location of a particle is smeared over a certain range. This uncertainty is reflected in the wave function that represents the particle.

Thus, quantum mechanics encompasses certain limits that are not found in classical mechanics. Classically, one can calculate that an electron is located at some point in space. Quantum mechanically, one can assert only that the position of an electron is spread out over a certain interval. Classically, one can estimate the speed of an electron. In quantum mechanics, an electron's velocity also is considered to be uncertain.

The only way that the position or velocity of a particle can be ascertained, in the quantum scenario, is through measurement. Quantum measurement is of a radically different character than classical measurement. In the quantum theory of scientific observation, the observer alters the nature of the wave function that represents the particle. Initially, the wave function cannot be said to have any specific value of the quantity that is being observed. Immediately upon measurement, the wave function takes on a specific value of the quantity that has been measured. When that occurs, the wave function is said to collapse. After the wave function collapses, it can be said to have the value of the quantity in question only for a brief interval after the measurement has taken place. If a measurement of another parameter occurs, then the wave function collapses again to a value of this second quantity.

We have already discussed that process in connection with the double-slit experiment (Chap. 4), in which the electron may pass through either of two slits. The wave function embodies both possibilities; it has an uncertainty. Only when one makes a specific measurement of the position of the electron can the electron be said to have passed through one slit or the other.

Let's consider what happens to an electron when both its position *and* its velocity are measured. Before those quantities are measured, the electron wave function is said to be in an indefinite (smeared) state of both position and velocity. Suppose a device now measures the position of the electron: The electron collapses into a state of definite position. One can now state with certainty where the electron is. Suppose a second device reads off the electron's velocity: The instrument would force the electron wave function into a state of definite velocity. The electron's position would again be uncertain. Neither quantity, position, nor velocity, could be measured simultaneously.

We note the strong effect that the scientific observer has on an experiment. In the standard (Copenhagen) interpretation of quantum mechanics, the ideas of definite position and velocity are curiously linked to human measurement. Without observation, no particle can be said to have specific values of its physical parameters. It seems as if quantum mechanics provides an answer to a hackneyed philosophical question: If a tree falls in a forest where there are no listeners, can it be said to make a sound? The Copenhagen interpretation of quantum mechanics implies that the answer is no: One cannot say that there is a sound. Only when there is measurement, can the falling tree be said to make a sound at a definite time.

It is in the extension of quantum mechanics to objects of human scale that the paradoxical nature of a collapsing wave function becomes apparent. It is hard to imagine that this collapse, and hence the assumption of fixed physical values such as speed and location, can depend so strongly on the existence of human beings. It seems to be a rather anthropocentric interpretation of physical reality. Yet that is the gospel of quantum mechanics, and it has been experimentally verified thousands of times!

Let us consider a famous case in which the Copenhagen interpretation of quantum mechanics is applied to a situation of macroscopic (ordinary) scale and yields astonishing results. This example, the Schrödinger cat paradox, is a thought experiment proposed by one of the developers of quantum theory.[3] In this imagined experiment, a cat is placed in a large cage. Next to the cat is an apparatus which, if activated, causes certain death of the creature. Triggering the apparatus is a particle detector that measures the spin of an electron. If a particular electron's spin is up, the apparatus is activated and the cat dies. If the spin is down, the apparatus is not triggered and the cat survives. The procedure is simple and unambiguous.

The whole experiment is covered and removed from view. An electron is sent into the machine. Since the electron's spin has not been measured before the electron enters the machine, it cannot be said to have a definite value. Thus, the electron's wave function is a mixed state of up and down.

After the electron has entered the detector, the detector shows a reading. However, since one cannot know what the reading is, since there has not been a human observation, it cannot be said to have a specific value. Therefore, the detector is in a mixed state: Its wave function is a mixture of up and down signals.

Finally, we turn to the question of the cat itself. Since the apparatus designed to perform the function of killing the cat is hooked to the detector, its wave function, like that of the detector, also must be in a mixed state. So the cat must be in a hybrid condition: half alive and half dead. More precisely, its wave function is a mixture of a dead cat quantum state and a live cat quantum state!

Nevertheless, if one were to remove the cover, one would see either a dead cat or a live cat; one would not see a mixed state. The Copenhagen interpretation of this is that the wave function of the cat *collapses* to either a live state or a dead state at precisely the moment when the cover is taken away. Human measurement directly causes the collapse to occur. Hence, scientific observation causes a cat that is in a condition of indeterminate existence to become either alive or dead!

It is not too hard to see why many quantum theorists have found Schrödinger's cat paradox to be disturbing. It is difficult to imagine how the values of physical quantities can be affected so strongly by the existence of an observer. We like to believe that there is an objective reality to the universe, one that transcends the realm of consciousness. Yet the Copenhagen interpretation tells us that consciousness plays a strong role in physics because it alters the nature of certain physical properties. As the cat paradox shows us, it can even transform an animal from something that is in a strange mixed state of being to a creature that is alive or dead—certainly a puzzling state of affairs!

The Many-Worlds Interpretation of Quantum Mechanics

In 1957, Hugh Everett, a Princeton University graduate student, published a thesis offering a highly exciting alternative to the Copenhagen interpretation. His idea, called the many-worlds interpretation of quantum mechanics, has generated a great deal of controversy. Most theorists have found his approach to be brilliant and compatible with observation but at the same time to be, as John Wheeler puts it, "carrying too great a load of metaphysical baggage." In other words, it seems too bizarre to be true!

Basically, Everett's idea is to remove the notion of a collapsing wave function from quantum mechanics. Instead of collapsing, the wave function is constantly splitting into myriads of copies of itself. Each copy corresponds to a valid picture of reality in one branch of the universe. In that manner, the universe is continually bifurcating.

Everett's model differs from the Copenhagen interpretation in its portrait of scientific measurement. In the many-worlds picture, any measurement of some quantity forces the observer into selecting one of the alternative branches of the universe. Parts of the wave function continue to exist in other worlds, but the observer no longer has access to them. By the selection process, the observer sets out irreversibly on one of many paths of time. A particular reality is chosen from all of the possible fates of the universe. In addition, numerous copies of the observer set out on other paths of time. Each of the observers remains unaware that the others exist.

Critics of the theory have argued that no one has observed the universe to split; that no observer has ever seen the process by which reality is selected from various possibilities. However, that point cannot be used as a successful argument against the theory. As Everett points out in his thesis:

> The whole issue of the transition from "possible" to "actual" is taken care of in the theory in a very simple way: There is no such transition, nor is such a transition necessary for the theory to be in accord with our experience. From the viewpoint of the theory *all* elements of a superposition (all "branches") are "actual," none any more "real" than the rest. It is unnecessary to suppose that all but one are somehow destroyed, since all the separate elements of a superposition individually obey the wave equation with complete indifference to the presence or absence of any other elements. This total lack of effect of one branch on another also implies that no observer will ever be aware of any "splitting" process.[4]

Let's see how the many-worlds interpretation provides an alternative explanation of the Schrödinger cat paradox. In Everett's model, there is no such thing as a mixed state; therefore, at all times, the cat is either alive or dead. In one branch of the universe, the cat has been killed; in the other branch, the cat has been spared. By making an observation, an observer finds himself in either one branch or the other, depending on the experimental outcome. Time has forked, with the observer setting out on one of the paths.

Note that there now are two copies of the human observer. The replica experiences the alternative outcome of the experiment. To him, this alternative result is the one reality. There is no way for the two copies of the observer to make contact. Each is unaware of the split that has taken place.

A similar branching of time can account for the results of the double-slit experiment. In one branch of the universe, an electron travels through the first slit; in the other, the electron travels through the second. By making an observation of the position of the electron, an experimenter splits into two alternative realities. The observer's "life line" forks, with the two copies perceiving two different versions of the course of events. One of the copies is certain that the electron has traveled through the first opening. The other is equally sure that the electron has traveled through the second. Each is correct with regard to the chronicle of events of the particular fork of the universe.

When we look back at the course of the history of the universe, what we perceive, according to this theory, is the end result of a myriad of choices among scores of possibilities. One might wonder why, given the large number of choices, a particular path has been chosen. Why has the universe evolved through one set of events rather than another?

In order to answer that question, one might postulate the use of the anthropic principle as a type of selection mechanism. Given the numerous possibilities of paths that our universe could have taken, it has traveled along one that leads to the emergence of intelligent life. Many of the other copies of the universe have not produced the requisite conditions for the evolution of conscious observers. These universes are simply not "aware" of themselves. Hence, a Darwinian selection process weeds out versions of the universe in which conscious life can evolve.

The many-worlds theory has not won broad acceptance among the mainstream community of physicists. In spite of the philosophical problems posed by the Copenhagen interpretation, the majority of theorists find the difficulties raised by the many-worlds interpretation to be far more formidable. The theoretical physicist Bryce De Witt, who has been among the leading advocates of Everett's approach, has expressed an appreciation of the problematic aspects of this interpretation:

> I still recall vividly the shock I experienced on first encountering this multiworld concept. The idea of 10^{100+} slightly imperfect copies of oneself all constantly splitting into further copies, which ultimately become unrecognizable, is not easy to reconcile with common sense.[5]

Since it is impossible to prove or disprove the existence of forked time, the many-worlds concept remains an interesting curiosity. It is truly a matter of taste whether or not one accepts the notions of the theory. Most physicists have chosen to continue to follow the Copenhagen interpretation in spite of the puzzling nature of wave function collapse. Only a few have continued to embrace the many-worlds interpretation.

The notion of time as a "garden of forked paths" is certainly a fascinating one. Could it be true, as Borges has written, that "time forks perpetually toward innumerable futures?" Is the world engaged in a process of continual bifurcation?

It is interesting to consider the possibility that alternative versions of reality exist in other branches of the universe. Unfortunately, since we have no access to them, these alternative pictures remain ghosts of what may have been. Like the wanderers in a labyrinth, we cannot say with certainty what lies in the countless inaccessible corridors that border our paths. As with many questions about the nature of time, we can only speculate.

References

1. Jorge Luis Borges, "The Garden of Forking Paths," in *Labyrinths,* New Directions Publishers, New York, 1962.
2. Ibid.
3. John Gribben, *In Search of Schrödinger's Cat,* Bantam, New York, 1984.
4. H. Everett III, *Rev. Mod. Phys.* **38**:453 (1957).
5. Bryce De Witt, "Quantum Mechanics and Physical Reality," *Physics Today,* September 1970, Vol. 23, p. 4.

7
Cracks in the Pavement

Time Atoms

In this chapter we turn to the question of the divisibility of time. It is interesting to examine the issue of whether or not time can be subdivided indefinitely. Is time smooth and steady, or does it have a smallest constituent? In the former case time would be like a continuous substance without gaps in its flow; in the latter, it would be formed of particles in much the same way that a stream is formed of rapidly moving tiny droplets.

Gazing at a stream of water cascading down a rock, we perceive a continuous gushing flow as it tumbles over the stony formation. If there is enough water, this constant movement appears to us as a solid white wall of fluid enveloping the stream bed. We see no breaks in the spray, no individual particles forming the greater flow.

On the other hand, if we view water dripping from a tap, we have quite a different picture: We can clearly make out individual water droplets, particles that make up the flow, each one following its own path as it falls from the faucet into the sink. Between the dripping of each drop there are gaps, both in space and in time, creating a continuous steady rhythm.

It is clear, of course, that there is no conflict between the images of a waterfall and a dripping tap. The only difference between them is that, in the former case, because of rapid motion, we cannot perceive the fact that the stream of water can be subdivided into its particle constituents and, in the latter case, the motion is slow enough that the droplets can be detected. There is no physical difference between the two flows; there is only a difference in our perception.

If we waited until nightfall and then aimed a strobe light at the cascading water, the true picture would become obvious. In that case, the flashing light would illuminate a series of individual droplets as they fell to the stream. The strobe light would break the continuity of our visual picture; it would reveal the true discrete nature of the fluid and thus show that the white curtain of water was an illusion. Our mind cannot distinguish particles when they are moving rapidly; it fills in the gaps and thus creates an illusionary continuity.

Traditionally, time has been considered to be a continuous quantity and has often been described as a steady, uninterrupted flow like a rapidly moving current of water. Yet, could it be that, like the waterfall, time's current can be broken up into individual droplets? One might argue, however, that our perception of time does not seem to contain any gaps. There seem to be no breaks in our states of wakeful awareness, because our streams of conscious thought proceed continuously. How can we imagine schisms in our very being? If time particles do exist, why do we perceive temporal continuity?

We might think of an analogy with motion pictures. When we go to the cinema, we observe a continuous image of sights and sounds, yet the images are produced by a series of still shots. A reel of movie film is simply a set of ordinary photographs intended to imitate continuous motion. At the normal rate of cinematic projection, our minds can fill in the spaces between these still photographs; if the movie projector is slowed down, we can readily notice the gaps. That is why the motions of actors in old silent films are jerky and uneven; the films are often shown at the wrong speed.

Our minds, then, have tremendous capacity to "fill in the gaps" between still images. Therefore, could it be possible that we are always doing that and time itself is discrete? Could the continuity of time be a product of the imagination? In that case, though psychological time would be continuous, real physical time would be composed of discrete units.

Physics generally treats time as a continuous variable. Like position or velocity, time is thought of as a quantity that can take on any of a smooth range of values. That is true whether time is considered to be circular or linear, static or relativistic.

Over the past three-quarters of a century, starting with the appearance of quantum mechanics, the idea that a physical quantity can be discrete has become more widely accepted. It is now commonplace to think of quantized (discrete) position, velocity, charge, momentum, and energy. Similarly, it is well known that all matter is made of atoms and that atoms, in turn, are made up of subatomic particles. Thus, all substances are composed of combinations of small constituents such as protons, neutrons, electrons, and neutrinos. Protons and neutrons themselves are considered to be composed of smaller units called quarks. Finally, quarks, electrons, photons, and neutrinos are thought to be basic and indivisible.

So, for instance, electric charge can be found in nature only in multiples of a smallest basic unit. Every bolt of lightning, every spark emanating from a battery, and every electric current from a wall outlet contains a quantity of electricity that is a product of this smallest unit of charge, that of one electron. That is why we say that electric charge is quantized.

Considering all that, one might wonder if time is composed of small particles. The logical extension of quantized matter would seem to be quantized time; therefore, one might postulate the existence of "time atoms" as the smallest units of time.

In spite of the apparent novelty of the notion, the idea of temporal atomism has a long history. A Buddhist sect called the Sautrankitas, which appeared in about the third century B.C., held that the universe is recreated every instant. They supposed that the continuity of time is merely an illusion. A number of Greek philosophers of the Hellenic period also subscribed to that view and argued vehemently that time was atomistic.

Zeno's Paradoxes

Not all of the early classical philosophers favored the notion of discrete time. The earliest recorded argument *against* the idea is that of the controversial philosopher Zeno of Elea.[1] Zeno, who lived in southern Italy in about the fifth century B.C., constructed four paradoxes of time and motion. The ingenuity of the paradoxes is demonstrated by the fact that they are still debated today.

Two of the paradoxes were composed to demonstrate the impossibility of time atoms. They are commonly referred to as the stadium paradox and the arrow paradox. In them Zeno asks us to imagine that time is discrete and then attempts to show us that the supposition leads to a contradiction.

The stadium paradox can be expressed in the following manner: Imagine a large sports arena that is filled with runners standing on a track. The track is marked off in uniform intervals of, let's say, one meter between successive marks. Thus, the first line on the ground would be marked 1 m, the second, 2 m; and so on. The surface of the stadium would look a bit like an American football field.

Imagine that two teams of runners are standing on the field with one runner per team on each meter mark. Let's call the teams the Athenians, team A, and the Brindisians, team B. Initially the first Athenian runner and the first Brindisian runner would stand on the 1-m mark, runner 2 from each team would stand on the 2-m mark, and so on. In that manner, every team member would be standing on the interval marker corresponding to his own number.

Let's now suppose that time, for the Athenians and Brindisians, is discrete. In other words, there is a smallest time interval such that no motion, thought, or action whatsoever could take place in less than that interval. Consider, for instance, that this interval, the time atom, is exactly one second. Then, for the members of teams A and B, nothing could occur in less than one second.

Now let's suppose that the runners of Athens are moving to the left along the track at the rate of one point per instant (1-m/second). Thus, after 1 second, Athenian runner 2 is standing on the stadium's 1-m mark, runner 3 on the 2-m mark, runner 4 on the 3-m mark, and so on. Here we use the word "instant" to refer to what we have taken to be the time atom, which in this case is 1 second. That corresponds to the assumption that time is discrete.

Let's also imagine that the runners of Brindisi are moving to the right at the rate of one point per instant. Therefore, after 1 second, Brindisian runner 1 is standing on stadium marker 2, runner 2 on marker 3, and so on. We also note that the first Brindisian is sharing the same marker as the third Athenian, the second Brindisian with the fourth Athenian, etc.

The following tables represent the positions of the first three runners from each team (labeled A for Athenian and B for Brindisian), relative to the stadium markers (labeled S), before and after the first instant:

Table 1.

	Before			After			
A	1	2	3	1	2	3	
S	1	2	3		1	2	3
B	1	2	3			1	2	3

Since we suppose that the instant in time (second) is indivisible, we cannot speak of a time *between* the two events. However, let's consider the following moment that this configuration must pass through, half-way between the before and after pictures:

Table 2.

		Intermediate moment				
A	1		2		3	
S		1		2		3
B			1		2		3

Note that, during this intermediate moment (½ second), A2 and B1 are aligned. This is a distinct recorded event, one to which we can certainly assign a time label. Therefore, one can refer to this time as a subdivision of one instant. This contradicts the idea that instants are indivisible, leading to a paradox.

In other words, according to the premise of Zeno's stadium paradox, no event can take place in less than one instant (second). Yet according to its conclusion, we can imagine an event (as in Table 2) taking place in less than one instant (½-second). Thus, this conclusion contradicts the premise that there exists a smallest unit of time. Hence, according to this argument, discrete time is impossible.

Zeno's arrow paradox is even more subtle. It is concerned with the reality of motion, and it can be stated as follows: Consider time to be discrete, so that there is a smallest time interval for motion. No motion can take place in less time than one instant. Now imagine an arrow, which we shall call arrow A, that is traveling through the air. At each instant, we can take a picture of the arrow and show that it is not moving. Its motion must occur *between* the instants, not *during* the instants, since one instant is the smallest time interval for an event to occur.

Now picture another arrow, which we shall call arrow B, that does not move at all. Compare a picture of arrow B, during a particular instant, with arrow A, during the same instant. Both arrows appear in the picture to be perfectly still. There is no way to distinguish arrow A from arrow B. However, when the next tick of the clock signals the coming of the next instant, arrow A will have moved, whereas arrow B will have remained still. Hence we see a paradox: During one instant, there is no difference between the appearance of the two arrows, yet one arrow acts in a completely different manner than the other!

The arrow puzzle is much easier to address than the stadium question. The answer to this paradox, as the philosopher Adolf Gruenbaum has pointed out, lies in the fact that mobility can arise from a series of immobilities. One can define motion as a series of successive events in which an object's location is altered. Thus, movement can stem from a collection of instances in which an object is at rest.[2]

One can draw an analogy with a work of art. Suppose one were to divide the *Mona Lisa* into its smallest components. None of these components would be recognized as belonging to Da Vinci's masterpiece, yet put them all together and the portrait is instantly recognizable. Similarly, even if, for each instant, the arrow is at rest, it can still exhibit an overall motion seen by putting the pieces together. Thus, Zeno's arrow paradox does not represent a true contradiction.

In spite of the questions posed by Zeno's paradoxes, philosophical and scientific interest in time atoms continued. During the Middle Ages, Maimonides, the renowned twelfth-century Jewish scholar, spoke of temporal atomism in his influential *Guide for the Perplexed*. In this work he explained that "time is composed of time atoms, i.e. of many parts, which on account of their short duration cannot be divided." Maimonides estimated that these time atoms have an extremely short duration:

> An hour is divided into sixty minutes, the minute into sixty seconds, the second into sixty parts, and so on; at last after ten or more successive divisions by sixty, time-elements are obtained which are not subjected to division, and in fact are indivisible.

To link together the individual moments, Maimonides looked to the providence of God. God is continually recreating the universe, he thought. If God were to disappear, the universe would freeze into a static state and no motion would be possible. Only the holy spirit can bridge these gaps in time.

In several religious and mystical belief systems, temporal atomism played an important role in explaining how an external spirit or force was necessary for the world to continue. If the world must be recreated every moment, divine providence is needed for the task. Instead of just one creation, an infinite number of creations take place, requiring that God play an active part in the world.

The writer Borges was intrigued by the concept. In his short story, "Tlön, Uqbar, Orbis Tertius," he paints a picture of a society which believes in discrete time. According to the philosophy of this fictional civilization, the world is recreated and destroyed every instant. Thus, if an object were placed somewhere at a given time, and were then viewed an instant later, it would be considered by the members of this fictional society to be a completely different object. For instance, if a penny were dropped on the ground by a merchant and then found again by a small child, the child's penny would be considered to be a different object than the merchant's penny. So, by implication, the only link between one moment and the next would come from some outside divine force. The fact that an object had temporal duration would be considered a miracle, much as if a cloud instantly disappeared from one part of the sky and reappeared in another part.

Imagine living in such a world where each "time atom" constituted a separate reality! One's life would have no continuity and hence no meaning. Could a person living such a life be said to have a "soul?"

That is the question asked by Oliver Sacks in *The Man Who Mistook His Wife for a Hat*. His patients with Korsakov's syndrome, such as Jimmie and Mr. Thompson, no longer can remember the past and live with only the knowledge of the immediate present. So, like the dwellers of Tlön, in some sense Sacks's patients dwell in a world in which time is broken up into pieces. In his book, Sacks wonders if his patients could be said to possess human psyches, since their inner lives are so detached from temporal continuity.

It seems that one's humanity, one's consciousness, and one's soul depend on the fact that time is continuous and that history exists. Without these bridges linking the particles of time, our lives would be completely shattered. Luckily, our minds have the capacity to create these links!

In a certain sense, our mental worlds are created and destroyed at every instant. Our sensory perception and mental processing occur at discrete intervals. Our minds link these disparate images together into one complete picture. When that mechanism fails us, the effects are immediately noticeable.

There is some evidence that neural processing occurs discretely, in units of $1/10$ second. Sensory information input at a faster rate simply cannot be absorbed into the brain. For example, certain psychedelic drugs are known to slow down the rate of visual processing and cause a user to experience discrete, choppy images. Under the influence of these substances, objects seem to be fixed in the visual field and appear as streaks. Fortunately, under normal circumstances, our minds link up our discrete visual images into one continuous stream of images.

Discrete Time in Physics

Unlike temporal atomism in mental processes, there is virtually no scientific evidence for physical time atoms. There are, however, speculative models of particle interactions which assume that time has a smallest unit. The so-called lattice gauge theories use discrete time as a calculational tool. In these constructs, it is assumed that particle interactions occur at fixed intervals. No event in the particle world can take place during a time less than the minimum interval. These models are proposed, not as ways of representing the real physical universe but as shortcuts to solve outstanding problems of physics that cannot be addressed with continuous versions of time.

In another example of the interest in time atoms, the Nobel prize-winning physicist T. D. Lee has recently developed a version of quantum mechanics in which time is discrete. In his paper, "Can Time be a Discrete Dynamical Variable," he discusses how quantum mechanics can be modified for that purpose. He concludes that such a notion is workable and may even be necessary.[3]

Cellular automata models of the cosmos, such as Fredkin's picture of the universe as a computer, also assume that the dynamics of the world are discontinuous. In Chap. 3 we outlined some of those theories. Unlike Newtonian mechanics, the interactions of the models take place during finite time steps. Between each two steps, the automata, like the arrow in Zeno's paradox, are clearly at rest. Therefore, the models provide us with pictures of temporal atomicity.

A simple analogy can illustrate the difference between continuous and discrete time models of physics. Imagine that two lawyers, living at opposite ends of a city arrange to meet at a certain time to have a discussion. Since Doris, the first lawyer, lives in the west end, Cloris, the second lawyer, must travel from the east end to reach her. Suppose that Cloris decides to drive. It takes 15 minutes by car to reach Doris; therefore, if Cloris sets out between 2:00 and 3:00, she should reach Doris between 2:15 and 3:15. In that case, the "interaction" between Doris and Cloris could take place any time during that hour-long period. Hence, the time of interaction would be a *continuous* quantity because it could be any one of a continuous range of values.

On the other hand, suppose that Cloris decides to take the bus. Buses run from the east end to the west end every 15 minutes and take 15 minutes to complete their route. If Cloris decided to leave the east end between 2:00 and 3:00, she would have to take either the 2:00, 2:15, 2:30, 2:45, or 3:00 bus, and would therefore arrive in the west end to meet Doris at 2:15, 2:30, 2:45, 3:00, or 3:15. She could not meet Doris at any time in between, for no bus could take her there at any intermediate time. So in this case the time of interaction between Doris and Cloris would be a *discrete* quantity because it could be only one of a finite set of values.

We can now understand the way in which the continuous time of the car trip is qualitatively distinct from the discrete time of the bus route. By analogy, we can see the difference between continuous and discrete models of physical interactions. Discrete physics requires that communication between particles occurs only after fixed intervals, whereas continuous physics has no such restriction.

Several decades ago, there was speculation among physicists that a quantized unit of time, called the chronon, exists. The value of this interval would be the time for a particle of light, a photon, considered to be the fastest particle, to pass an electron, considered at that time to be the smallest particle of nonzero diameter. Since the speed of light in a vacuum is 10^8 m/second and the effective diameter of an electron is roughly 10^{-15} m/second, this time unit would be of the order of 10^{-23} second. No interactions could take place at a shorter time. The chronon would form the basic unit of all temporal measurement.

With the increased understanding of particle physics of recent years, natural units of time that are even much smaller than the minuscule chronon have been proposed. One unit of temporal duration that has come into increasing use in cosmology is called the Planck time. This interval, roughly 10^{-43} second in magnitude, is the amount of time for light to travel a distance known as the Planck length. The importance of the Planck length is that it is the smallest distance for which Einstein's theory of general relativity is valid as a classical theory of gravity. Dynamics taking place in smaller intervals must be described by a quantum theory of gravity which remains a highly speculative notion.

Aside from these highly speculative notions, most physicists believe that time is a continuous quantity. No one has shown conclusively that time atoms exist. In the absence of evidence to the contrary, one must assume that time is infinitely divisible.

References

1. G. J. Whitrow, *The Natural Philosophy of Time*, Harper and Row, New York, 1963.

2. Adolf Gruenbaum, *Modern Science and Zeno's Paradoxes*, Wesleyan University Press, Middletown, Conn., 1967.

3. T. D. Lee, "Can Time be a Discrete Variable?" *Phys. Lett.* **122B**:217 (1983).

8

The End of the Line

We have now completed our journey through the many different forms and facets of time. We have seen how time has been shaped into many molds by philosophers, historians, theologians, psychologists, and physicists. Although scholars from these separate fields have different backgrounds, priorities, and purposes, their views of time have many similarities. The viewpoints naturally fall into several different categories, as our discussion has indicated.

We have discussed the contrast between linear and circular time. Within the broad area of linear time, there are two different points of view: the pessimistic and optimistic philosophies. Our discussion has also included the notions of causality and synchronicity, focusing on the question of whether or not a time lag is needed for information exchange. We have seen how the flow of time can be affected by various external factors and have detailed the possibilities of time travel. The concept of forked time has been reviewed. Finally, we have shown how continuous time can be distinguished from atomistic or discrete time.

In all those cases, we have explored how the concepts originated as well as how they are viewed by contemporary scholars. Because time is a highly complex and elusive concept, ancient debates over its meaning and "shape" continue to be waged, now using the modern tools of scientific enquiry. It is entirely possible that time will never be fully understood. Truly, only time will tell.

Why do some thinkers favor certain models of time but reject others? That is a rather broad question, the answer to which clearly depends upon the discipline in question. A physicist might reject a certain temporal model because of certain theoretical considerations. Another scientist might point to experimental evidence which favors a particular picture of time. Finally, a theologian may show how one view of the world is consistent or inconsistent with Biblical teachings. Each of these thinkers might reach similar conclusions but arrive at them through radically different thought processes.

On the other hand, human beings have certain vital features in common: all are born, grow old, and die. This commonality of experience cannot help but be reflected in our attitudes toward time. For, in some sense, one's view of time is related to one's attitude toward death and finality. That is not to say that scientific experimentation is prejudiced by human fears and desires, but when conclusions are drawn from certain bodies of evidence, a large measure of interpretation must be included. That is the point at which aesthetic and psychological considerations can often enter the picture. One's attitude toward life and death can be consciously or subconsciously incorporated into one's choice of several alternative perspectives on some philosophical or scientific question.

Let us now examine how one's feelings toward human mortality can affect one's view of the structure of time. The existence of death presents the strongest influence that time has upon a human being. The philosopher Heidegger has stated that we perceive time only because we have to die. Naturally, one's perspective toward time is a function of one's attitude toward its ravage.

Static models of time are most attractive to those who are dismayed by thoughts of the finiteness of human existence. The steady-state and Newtonian portraits of the universe allow for the possibility that the human race will exist forever, and species immortality is somewhat of a consolation for those who are disturbed by death. Proponents of the steady-state theory of the universe cannot imagine that the universe has a beginning or end. Everything will be as it always was, they feel.

Oscillating cosmological models offer the reassuring view that nothing is lost in the end. Everything will return to its original state again and again, ad infinitum. As Nietzsche relates, immortality is guaranteed to us in the form of the eternal return. From earliest times, rituals and repetitive behavior have been used to ward off death and establish cultural longevity. The Indian concept of reincarnation allows for the soul to return in various guises. In all these cases, there is hope that renewal and recovery will follow periods of destruction and hopelessness. Here the perennial nature of agriculture and celestial motion serve as a guide for a way to restore youth and vigor through eternal repetition.

The linear, downward slope model of time appeals to a different sort of thinker. Decay is looked upon as a necessary process, and death and destruction are seen as inevitable and irreversible. The law of entropy encodes decay into the structure of physical theory. A similar purpose is served by the theological adaptation of Armageddon as the end goal of human history. In both cases it is believed that humankind cannot maintain itself and its society indefinitely. The only hope is that a miracle will take place. In lieu of this possibility, downward slope philosophers suggest that one should accept one's lot and make the best use of the present moment.

Upward slope philosophers of linear time are a far more optimistic group. Progress is seen as the hallmark of social and political maturation. Although human biology has evolved from inferior forms, human society continues to develop for the best. Upward slope model physicists speak of complexity appearing from chaos and order stemming from random behavior. It is hoped that humanity will overcome death and despair by exploiting technological achievements, expanding to other parts of the universe, embracing social unity, or by pursuing other means for growth. The common dream is one of utopia.

Another way to deal with the devastation that time enacts is to search for means of escape. This escape can take place through the use of drugs, the timelessness of sleep, or retreat into the deepest recesses of one's mind. One might also consider the possibility of time travel or slowing or hastening the flow of time. All those avenues open up the attractive prospect of stepping out from the flowing river of time.

Finally, let us consider the many-universes theory that time is constantly forking. Even this abstract theory offers reassurances with regard to the continuity of human life. In fact, this picture of the cosmos guarantees immortality for everyone! Since according to this model the universe continually bifurcates, if one dies in one branch of the universe, one can live in others. There are an infinite number of timelines along which death can be postponed indefinitely.

For those of us who believe in the power of time but do not fully understand it, there can be little true satisfaction with any of these models in particular. The complete nature of time seems to be unknowable, with many pieces of its puzzle still undiscovered. As Borges said in one of his last interviews, "Time is the one essential mystery."

Time is the path that leads us toward our goals. It is the substance of our lives, yet in the end it cannot fail to betray us, as we are brought to the last precipice. The same hands that mold us are the hands that push us over the edge. Hermann Hesse has described the strange way that time creates and destroys us:

> No permanence is ours; we are a wave
> That flows to fit whatever form it finds:
> Through day or night, cathedral or the cave
> We pass forever, craving form that binds.
>
> Mold after mold we fill and never rest,
> We find no home where joy or grief runs deep.

We move, we are the everlasting guest.
No field nor plow is ours; we do not reap.

What God would make of us remains unknown:
He plays; we are the clay to his desire.
Plastic and mute, we neither laugh nor groan;
He kneads, but never gives us to the fire.

To stiffen into stone, to persevere!
We long forever for the right to stay.
But all that ever stays with us is fear,
And we shall never rest upon our way.[1]

Reference

1. Hermann Hesse, *Magister Ludi — The Glass Bead Game*, Bantam Books, New York, 1969.

Index

About the Author

Paul Halpern received his doctorate in theoretical physics from the State University of New York at Stony Brook in May of 1987. He has written numerous articles on chaos, nonlinear dynamics, and the structure of space and time for such prestigious publications as *The American Journal of Physics* and *Physical Review*. In 1987 he was invited to be a visiting assistant professor of physics at Hamilton College in Clinton, New York. He is currently an assistant professor of mathematics and physics at the Philadelphia College of Pharmacy and Science.